Functional Composite Materials: Manufacturing Technology and Experimental Application

Edited by

C. Samson Jerold Samuel
Department of Mechanical Engineering,
Sri Krishna College of Engineering and Technology,
Kuniamuthur, Coimbatore
India

M. Suresh
Department of Robotics and Automation Engineering,
PSG College of Technology,
Peelamedu, Coimbatore
India

&

S. Gnanasekaran
Department of Mechanical Engineering,
Sri Shakthi Institute of Engineering and Technology,
Coimbatore,
India

Functional Composite Materials: Manufacturing Technology and Experimental Applicationson Diabetes

Edited: C. Samson Jerold Samuel, M. Suresh and S. Gnanasekaran

ISBN (Online): 978-981-5039-89-4

ISBN (Print): 978-981-5039-90-0

ISBN (Paperback): 978-981-5039-91-7

Bentham Science Publishers Pte. Ltd.
80 Robinson Road #02-00
Singapore 068898
Singapore
Email: subscriptions@benthamscience.net

BENTHAM SCIENCE

CONTENTS

FOREWORD

The book Functional Composite Materials: Manufacturing Technology and Experimental Applications presents research work on polymer matrix composites, metal matrix composites, metal and composite machinability studies, and resistance spot welding of the sheets.

This volume contains significant original and theoretical experimental results that make use of problem-solving technologies that are often unfamiliar to the general reader. This book's review and research papers present novel applications of experimental techniques and analyses of functional composite materials, indicating the need for novel analytical techniques.

N.L.Parthasarathi
Scientific Officer
MFTS/MDTD/MEG/MMG
IGCAR Kalpakkam,
India

PREFACE

The main attentions in this collection of scientific book chapters is focused on recent theoretical and practical advancements in functional composite materials related to mechanical and wear properties, microstructural characterization, and advanced manufacturing techniques used in aerospace, automobile, marine and defense industries.

This book highlights the advancements in the functional composites, metal matrix composites, polymer matrix composites, machinability studies of metals and composites using CNC turning process, electrical discharge machining, end milling, abrasive jet machining, electro chemical machining, additive manufacturing of products, joining of metals using resistance spot welding and carbon aerogels doped with resorcinol formaldehyde. We anticipate that this book will be of significant interest to the industrialists working on the basic issues surrounding the functional composites and manufacturing technology, as well as those who work in the industry in applied problems such as processing and applications. Because of multidisciplinary nature of the research, this book will attract a broad audience, including materials scientists, chemists, physicists, manufacturing and chemical engineers and processing specialists who are involved and interested in the future frontiers of blends.

The editors are grateful to their management, institution, family members and effort of theauthors who have contributed the book chapters. Additionally, without the tireless effort of the publication staff of Bentham Science Publishers, this book would have not been possible.

C. Samson Jerold Samuel
Department of Mechanical Engineering,
Sri Krishna College of Engineering and Technology,
Kuniamuthur, Coimbatore
India

M. Suresh
Department of Robotics and Automation Engineering,
PSG College of Technology,
Peelamedu, Coimbatore
India

S. Gnanasekaran
Sri Shakthi Institute of Engineering and Technology,
Coimbatore,
India

List of Contributors

Ajith Arul Daniel	Vels Institute of Science, Technology & Advanced Studies, Chennai 600117, Tamil Nadu, India
Amuthakkannan P.	PSR Engineering College, Sivakasi, Tamillnadu, India – 626140
Arumugaprabu V.	Kalasalingam Academy of Research and Education, Krishnankoil, Tamillnadu, India – 626126
Arunprasath K.	PSN College of Engineering and Technology, Tirunelveli-627152, Tamilnadu, India
Ashok Kumar	Tezpur University, Tezpur, Napaam-784028, Assam, India
AshokaVarthanan P.	Sri Krishna College of Engineering and Technology, Coimbatore, Tamilnadu, India - 641008
Deepak Joel Johnson R.	Saveetha Institute of Medical and Technical Sciences, Chennai, Tamillnadu, India– 600077
Ganesh N.	K Ramakrishnan College of Engineering, Samayapuram, Trichy, Tamilnadu621112, India
Ganeshkumar S.	Sri Eshwar College of Engineering, Coimbatore, Tamil Nadu, India
Gnanasekaran S.	Sri Sakthi Institute of Engineering and Technology, Coimbatore, TamilNadu, India
Jafrey Daniel James D.	K. Ramakrishnancollege of Engineering, Trichy, Tamil Nadu, India
Jeyakumar R.	Sri Krishna College of Engineering and Technology, Coimbatore, Tamilnadu, India - 641008
KarthikPandiyan G.	Sri Vidya College of Engineering and Technology, Virudhunagar, Tamil Nadu, India
Kalirasu S.	Sethu Institute of Technology, Pulloor, Kariapatti - 626115, Virudhunagar District, Tamilnadu, India
Madhankumar S.	Sri Krishna College of Engineering and Technology, Coimbatore, India-641008
Manikandan A.	Lord Jegannath College of Engineering and Technology, Nagercoil, India
Muthiah A.	P.S.R. Engineering College, Sivskasi, India
Muthukumaran Ramasamy	CMR Institute of Technology, Hyderabad, Telangana 501401, India
Nallusamy S.	Jadavpur University, Kolkata-700032, India
Nithya M.	CMR Institute of Technology, Hyderabad, Telangana 501401, India
Paulraj G.	SRM TRP Engineering College, Trichy, Tamilnadu, India 621105
Pitchipoo P.	P.S.R. Engineering College, Sivskasi, India
Prabaharan T.	Mepcoschlenk Engineering college, Sivakasi, Tamil Nadu, India
Prabu Subramanian	SRM TRP Engineering College, Trichy, Tamilnadu, India 621105
Prakash P.	Karpagam Institute of Technology, Coimbatore, TamilNadu, India

Raja S.	Sri Krishna College of Engineering and Technology, Coimbatore, Tamilnadu, India - 641008
Rajakarunakaran S.	Ramco Institute of Technology, Rajapalayam, India
Ramesh Babu R.	Karpagam Institute of Technology, Coimbatore, TamilNadu, India
Rituraj Dutta	Tezpur University, Tezpur, Napaam-784028, Assam, India
Saiyath Ibrahim A.	Karpagam Institute of Technology, Coimbatore, Tamil Nadu, India
Samson Jerold Samuel Chelladurai	Sri Krishna College of Engineering and Technology, Coimbatore, Tamil Nadu, India
Selvan T. A.	Sri Krishna College of Engineering and Technology, Coimbatore, India-641008
Senthil Kumar K. M.	Kumaraguru College of Technology, Coimbatore, India-641049
Senthil Kumar S.	SRM TRP Engineering College, Trichy, Tamil nadu, India 621105
Sujatha K.	Dr. M.G.R. Educational and Research Institute, Chennai-600095, Tamilnadu, India
S. Sundar	Department of Adult, Continuing Education and Extension, Jadavpur University, Kolkata-700032, India Department of Mechanical Engineering, Dr. M.G.R. Educational and Research Institute, Chennai-600095, Tamilnadu, India
Thilak M.	SRM TRP Engineering College, Trichy, Tamilnadu, India 621105
Venkatesh S.	Sri Eshwar College of Engineering, Coimbatore, India
Vigneshkumar M.	Sri Krishna College of Engineering and Technology, Coimbatore, Tamilnadu, India - 641008
Vijayan S. N.	Karpagam Institute of Technology, Coimbatore, Tamil Nadu, India
Viswanathan A.	SNS College of Technology, Coimbatore, India – 641035

CHAPTER 1

Optimization of Machining Process Parameters to Calculate Surface Finish, Metal Removal Rate, and Cutting Force on AA6061-T6 Alloy

Karthik Pandiyan G.[1,*], **T. Prabaharan**[2] and **Jafrey Daniel James D.**[3]

[1] *Department of Mechanical Engineering, Sri Vidya College of Engineering and Technology, Virudhunagar, Tamil Nadu, India*

[2] *Department of Mechanical Engineering, Mepco Schlenk Engineering College, Sivakasi, Tamil Nadu, India*

[3] *Department of Mechanical Engineering, K.Ramakrishnan College of Engineering, Trichy, Tamil Nadu, India*

Abstract: In this study, CNC turning process parameters are optimized while machining AA6061 – T6 alloy. The process variables, namely, cutting spindle speed or spindle velocity (Cs), rate of feed (Fr), rate of the depth of cut (Dc), and nose radius (Nr), were optimized with output responses such as surface roughness (Ra), feed power (Fx), thrust power (Fy), cutting power (Fz), and removal rate of material (MRR). The optimized results showed higher MRR, Fy, and Fz and lower Fx and Ra were obtained at a cutting spindle speed or spindle velocity (Cs) of 100 rev/min, rate of feed (Fr) of 0.2 mm/min, depth of cut (Dc) of 1.5 mm/min, and nose radius (Nr) of 0.8 mm. TOPSIS and ANOVA were used to assess the experimental results to determine the most favourable factors that were superior and suitable for selecting the optimal variable condition.

Keywords: AA6061-T6 alloy, ANOVA, CNC Turning, Cutting power, Depth of the cut, Feed power, Material removal rate, Multiresponse Optimization, Nose radius, Rate of feed, Spindle Speed, Surface roughness, Thrust power, TOPSIS.

INTRODUCTION

Aluminium is ductile and has an FCC (Face Centred Cubic) structure. Al is a lighter metal with 2900 kg/m^3 as density. It has higher electrical and thermal conductivity due to its lower density. Various wrought aluminium products include mechanical finishes, chemical finishes, and coatings.

* **Corresponding author Karthik Pandiyan G.:** Department of Mechanical Engineering, Sri Vidya College of Engineering and Technology, Virudhunagar, Tamil Nadu, India; E-mail: karthik.pandiyan007@gmail.com

C. Samson Jerold Samuel, M. Suresh & S. Gnanasekaran (Eds.)

Mechanical finishes include cold finished, buffed, and textured coatings, whereas chemical finishes include etched, bright dipped, and chemical conversion coatings. Anodizing, painting, plating, and chemical conversion coatings are included in the coating category. Due to low density to high strength ratio, toughness, and corrosion resistance, aluminium alloys are mainly used in automobiles and aerospace Industries. Zn, Si, Mg, Cu, are major alloying elements that are classified into cast-based alloys and wrought-based alloys with heat-treatable and non-heat-treatable properties.

Chemical composition of AA6061-T6 (T6- heat-treated, rapid cooling and aged) is Al - 97.25%; Cu - 0.32%; Mg - 1.08%; Mn - 0.52%; V - 0.01%; Fe - 0.17%;Sic - 0.63%; Ti - 0.02%, which is a work-hardened alloy with a low density to high strength ratio, increased hardness, improved toughness, capability of welding, and greater resistance to corrosion. The strain hardening method is highly effective in improving the strength of materials.

Due to tempering, materials lose their strength but show increased ductility, high tool wear and require elevated cutting power, and fatigue life of the tool. This results in difficulty of machining. An attempt was made to analyse multi-response characteristics based optimization of AA6061-T6 alloy, utilizing the TOPSIS method. In the previous examination, optimization was carried out based on the Taguchi S/N ratio-based technique. It is used as a tool to ensure better quality and efficient yield production at minimum expenditure and optimizes only a single response characteristic.

Normally, the Taguchi method is used for a single objective. Larger criteria are used for MRR and smaller criteria are used for surface roughness. Multi-responses are more difficult in the Taguchi method and are undertaken for converting the multi-response to a single response. MCDM uses Taguchi S/N ratio, Taguchi-based GRA, VIKOR, TOPSIS, and Analytic Hierarchy Process to determine the weights of each response in Decision-making methods.

An investigation was undertaken into the turning process where the input variable varied with different levels of input variable and output variables achieved as Ra, MRR, FX, FY, and FZ. A lot of research is available on the use of Grey, TOPSIS, and FUZZY TOPSIS to solve multi-objective problems, of which TOPSIS was found the most effective in the choice of variables and machining processes. Optimization combining TOPSIS and Taguchi methods achieved the most favourable conditions with satisfactory turning operations. Evolutionary computational methods have evolved, such as RSM, GA, PSO, and NN. But TOPSIS is still highly capable of solving MCDM problems to optimize the most favourable result.

In this study, Ra, MRR, FX, FY, and FZ were considered to have equal weights. Optimization study was undertaken to achieve surface roughness and wear rate of the tool by employing RSM, GRA, and TOPSIS for prediction [1]. TOPSIS was used at the end milling process to achieve Ra, lower energy, reduce the tool's wear rate, and improve yield in the manufacturing process [2]. Bing *et al.* experimented with 7050-T7451 aluminium alloy in the milling process to attain enhanced surface features by the ANOVA technique while evaluating specific cutting energy [3]. Gopal *et al.* used Taguchi-based S/N ratio, the TOPSIS and GRA ranking methods of milling on a carbide tool to predict the best optimum responses like cutting forces, cutting temperature, and Ra. Machining studies were conducted in the end milling process of Mg-based Hybrid metal matrix composite materials by GRA, TOPSIS, and Taguchi techniques [4]. Al alloys with superior surface finish were used to achieve lower surface irregularity, greater fatigue time, higher strength, high rate of wear, corrosion, and creep resistance to produce optimum surface quality [5].

End milling of Al6061 alloy was performed to achieve the best set of variables for the cutting force, which achieved tool deflection and cutting power [6]. Micro-milling of Al6061 alloy was performed to achieve Non-Uniformity of Surface Generation to ensure low feed rate and high depth of cut [7]. Szymon *et al.* investigated micro-milling of alloy steel to achieve good surface cease and minimum vibrations during milling [8]. Wojciechowski *et al.* investigated ball milling of hardened steel and achieved decreased deflection of tool diameter and tool extension [9]. Malghan *et al.* used the milling of AA6061 alloy with Taguchi-based optimization technique to evaluate process variables for greater surface irregularity [10].

Pereira *et al.* used the milling of Al 7075 alloy through TOPSIS to achieve an elevated removal rate of material, roundness, cutting power, and minimized energy [11]. Nipanikar *et al.* observed the tool wear of Ti-6Al-4V in a MQL environment using GRA and TOPSIS methods [12]. MQL based turning process was examined to determine the Multi-Objective characteristics of tooltip interaction using the Grey-Taguchi Method [13]. Design and Modelling of Inconel 718 alloy were studied to evaluate Tool Wear rate utilizing MQL [14]. Machining of Al-6082 T6 was performed to understand the characteristics of surface roughness [15].

Milling of Aluminum Alloy by using both Cryogenic enhanced CO_2 and LN2 Coolants was studied to evaluate cutting temperature in varying environments to achieve surface roughness [16]. A review was conducted on Aluminium alloys for usage and properties [17]. Taguchi-Grey Relational Approach on End Milling was performed to determine Multiple Response Characteristics [18]. AWJ Cutting of

D2 Steel was used to predict multi-response characteristics on process parameters [19]. End Milling Parameters were studied in Cryogenic Cooling Conditions [20].

Karthik *et al.* studied machining parameters on AA6351 alloy using RSM [21]. Karthik *et al.* predicted surface roughness using a soft computing method ANN [22]. The study represented an attempt to investigate multi-response characteristics of Turning process variables in AA6061-T6 alloy by using the TOPSIS method

MATERIALS AND METHODS

Experiments were carried out at various levels like cutting spindle speed or spindle velocity (C_s), rate of feed (F_r), depth of cut(D_c), and Nose radius in the turning process. Experiments s(N_r) were conducted in BATLIBOI model on CNC turning equipment. Table **1** demonstrates the machining variables of the AA6061-T6 alloy. Experiments were conducted for three levels of cutting spindle speed or spindle velocity (C_s) at 100, 200, and 300 rev/min, three rates of feed (F_r) at 0.10, 0.15, and 0.20 mm/min, depth of cut (D_c) at 0.5,1.0, and 1.5 mm/min and Nose radius (N_r) of 0.4, 0.6, and 0.8 mm.

Aluminium 6061-T6 alloy specimen with 150 mm length and 25 mm diameter was used as the workpiece. Table **2** shows the elements and composition of the AA 6061-T6 alloy, while Table **3** exhibits the properties of the AA 6061-T6 alloy considered for the study.

TPUN110408 116 type tool holder was used with CNMG12408-PM carbide inserts made by ISCAR. Table **4** displays the technical specifications of the CNMG12408-PM cutting insert.

Ra, Material Removal Rate (MRR), and cutting forces were measured in each experiment. Fig. (**1**) shows the decision model of the turning process variables.

Table 1. Machining variables of AA6061-T6 alloy.

Cutting Spindle Speeds Or Spindle Velocity (C_s) (rev/min)	Rate of Feed (F_r) (mm/min)	Depth of Cut (D_c) (mm/min)	Nose Radius (N_r)(mm)
100,200,300	0.10,0.15,0.20	0.5,1.0,1.5	0.4 ,0.6 ,0.8

Table 2. AA 6061-T6 alloy Elements and its Composition.

Element	Al	Ti	Mg	V	Fe	Mn	Si	Cu
Weight%	97.25	0.02	1.08	0.01	0.17	0.52	0.63	0.32

Table 3. AA 6061-T6 alloy Properties.

Density	2698kg/m^3
Melting Point	660°C
Elastic Modulus	75GPa
Poisson's Ratio	0.3
Tensile Strength	550MPa
Yield Strength	500MPa
Percent Elongation	25%

Table 4. Specifications of CNMG 120408-PM cutting insert.

Rhombic Nose Angle (°)	80°
Length (L) (mm)	12.96
Width (IC) (mm)	12.70
Thickness(S) (mm)	4.76
Nose radius (RE) (mm)	0.80
Relief angle (°)	15°

Fig. (**2**) shows the pictures of the experimental setup, and Fig. (**3**) shows the Turning process of the workpiece.

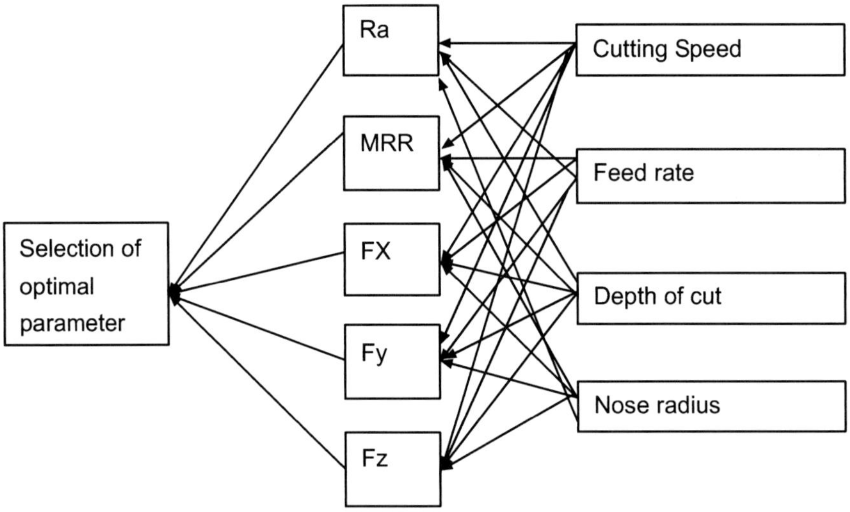

Fig. (1). Decision Model for Turning process variable.

Fig. (2). Experimental setup.

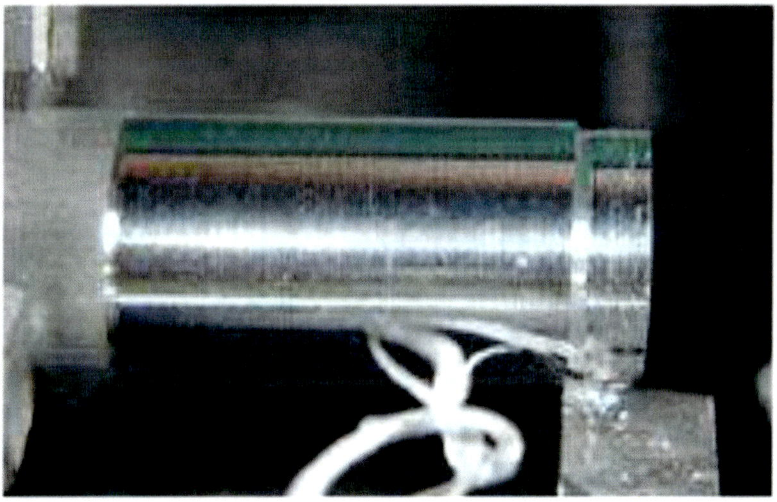

Fig. (3). Turning process of the workpiece.

Fig. (**4**) shows the uncoated CNMG 120408-PM grade-K20 carbide insert tool geometry, and Fig. (**5**) shows the force measure1ment setup, respectively.

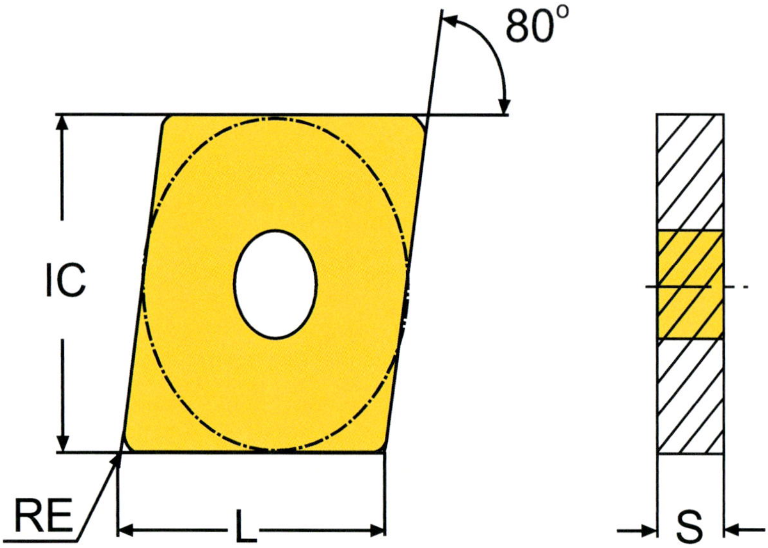

Fig. (4). Uncoated CNMG 120408-PM grade-K20 carbide insert geometry.

Fig. (5). Force Measurement Setup.

The experimental setup consisted of a three component piezoelectric type dynamometer used to measure force. The machined workpiece's surface roughness values were measured by a Mitutoyo surface tester (Make-Japan –Model SJ-301) with a cut-off length of 2.5 mm, speed of 0.5 mm/s, and a measuring range of 12.5 mm. Experiments were conducted with four variables at three levels.

Table **5** describes the Machining process variables at different levels. Table **6** shows the L27 orthogonal array with normalized values used to perform the experiments with each process variable on AA6061-T6 (150 mm (length) and 25 mm (diameter)). Ra, MRR, Fx, Fy, and Fz were considered output responses to locate the performance measures.

Table 5. Machining process variables with different levels.

Factors	Level 1	Level 2	Level 3
Cutting Spindle Speeds or (C_s) (rev/min)	100	200	300
Rate of Feed (F_r) (mm/min)	0.10	0.15	0.20
Depth of Cut (D_c) (mm/min)	0.5	1.0	1.5
Nose Radius(N_r) (mm)	0.4	0.6	0.8

Table 6. L27 orthogonal array with normalized values.

S. No.	Cutting Spindle Speeds or Spindle Velocity (C_s) (rev/min)	Rate of Feed (F_r) (mm/min)	Depth of Cut (D_c) (mm/min)	Nose Radius (N_r) (mm)	Normalised Matrix Value Rij				
					Ra, (μm)	MRR (mm^3/s)	FX (N)	FY (N)	FZ (N)
1	100	0.1	0.5	0.4	0.229	0.287	0.081	0.087	0.083
2	100	0.1	1	0.6	0.117	0.225	0.122	0.106	0.123
3	100	0.1	1.5	0.8	0.049	0.138	0.153	0.089	0.141
4	100	0.15	0.5	0.6	0.073	0.229	0.167	0.142	0.157
5	100	0.15	1	0.8	0.163	0.106	0.227	0.182	0.200
6	100	0.15	1.5	0.4	0.155	0.191	0.255	0.124	0.245
7	100	0.2	0.5	0.8	0.307	0.112	0.261	0.426	0.235
8	100	0.2	1	0.4	0.282	0.286	0.329	0.417	0.329
9	100	0.2	1.5	0.6	0.245	0.115	0.349	0.345	0.357
10	200	0.1	0.5	0.4	0.176	0.253	0.082	0.138	0.089
11	200	0.1	1	0.6	0.299	0.217	0.108	0.123	0.105
12	200	0.1	1.5	0.8	0.076	0.165	0.092	0.084	0.104
13	200	0.15	0.5	0.6	0.175	0.255	0.153	0.148	0.148
14	200	0.15	1	0.8	0.240	0.201	0.151	0.099	0.156
15	200	0.15	1.5	0.4	0.120	0.101	0.167	0.086	0.201
16	200	0.2	0.5	0.8	0.177	0.094	0.221	0.187	0.195
17	200	0.2	1	0.4	0.153	0.108	0.251	0.171	0.263

(Table 6) cont.....

18	200	0.2	1.5	0.6	0.228	0.136	0.254	0.184	0.272
19	300	0.1	0.5	0.4	0.049	0.204	0.078	0.108	0.075
20	300	0.1	1	0.6	0.140	0.295	0.095	0.142	0.098
21	300	0.1	1.5	0.8	0.036	0.226	0.110	0.138	0.108
22	300	0.15	0.5	0.6	0.205	0.136	0.130	0.138	0.128
23	300	0.15	1	0.8	0.295	0.200	0.161	0.206	0.163
24	300	0.15	1.5	0.4	0.277	0.256	0.189	0.126	0.198
25	300	0.2	0.5	0.8	0.241	0.115	0.145	0.189	0.154
26	300	0.2	1	0.4	0.022	0.116	0.220	0.241	0.208
27	300	0.2	1.5	0.6	0.136	0.127	0.245	0.173	0.263

Steps Involved in TOPSIS Method

Technique for Order Preference by Similarity to an Ideal Solution (TOPSIS) methodology was proposed by Hwang Yoon in 1981. This method uses positive ideal theoretical and negative ideal theoretical solution evaluation methods to calculate the coefficient value.

A Hypothetical solution for all experimental results consists of negative and positive ideal solutions. The obtained results should fall within the minimum and maximum values of the solution and the value of closeness coefficient between the minimum and maximum values.

TOPSIS is a ranking method to find the most favourable value from a set of results obtained from several experiments. The major principle of this method is that the final optimized result should be nearer to the positive values and distant from the negative values.

Fig. (**6**) shows the mean response value of the process variables. A specific set of output variables [Ra, MRR, Fx, Fy, and Fz] was given equal weight. TOPSIS method was mathematically used as follows [1, 4, 12, 19].

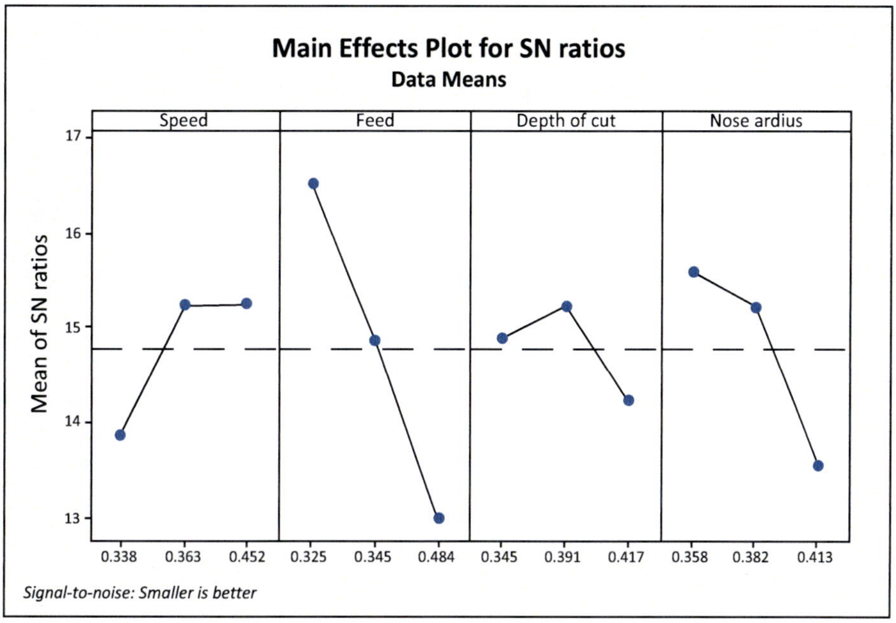

Fig. (6). Mean Response value of process variable.

Step-I: First of all, the values must be normalized to the matrix by the equation (1) as follows.

$$r_{ij} = \frac{a_{ij}}{\sqrt{\sum_{i=1}^{m} a_{ij}^2}} \tag{1}$$

Where i=1,2,…..27 j=1,2,…..5,

i denotes the number of observations

j denotes the number of output variables

a_{ij} denotes the normalized values of output variables

r_{ij} denotes the normalized matrix values of output variables

Step-II: The set of output variables [Ra, MRR, Fx, Fy, Fz] are given equal weight.

Step-III: The normalized weighted matrix with weights is determined using multiplication of the normalization matrix with its related weights in equation (2)

$$V_{ij} = W_i \times r_{ij} \quad \textbf{(2)}$$

Where i=1, 2, ...27 j=1, 2, ...5,

i denotes the number of observations

j denotes the number of output variables

V_{ij} denotes the normalized weighted matrix of output variables

W_i denotes the weight of the output variable

Step-IV: Determine the weight decision matrix where V+ is the maximum value of the particular output variable and V- is the minimum value of the particular output variable

$V^+ (V_1^+ V_2^+, \ldots\ldots\ldots V_n^+$ maximum values

$V^- (V_1^- V_2^-, \ldots\ldots\ldots V_n^-$ maximum values

Step-V: The below equation is used to calculate the values of S+ and S-

Where S+ denotes the best performance, S- denotes the worst performance

$$S_i^+ = \sqrt{\sum_{j=1}^{M}\left(V_{ij} - V_j^+\right)^2} \quad \textbf{(3)}$$

$$S_i^- = \sqrt{\sum_{j=1}^{M}\left(V_{ij} - V_j^-\right)^2} \quad \textbf{(4)}$$

where i = 1,2,....27.

Step-VI: The value of Closeness Co-efficient (CC_s) is determined by equation (5).

$$CC_s = \frac{S_i^-}{S_i^+ - S_i^-} \quad \textbf{(5)}$$

Step-VII: Ranking the CC_s result.

The highest CC_s value is close to the optimized value.

RESULTS AND DISCUSSION

Table **7** displays the L27 orthogonal array with the closeness coefficients standards and its position at every observation.

Table 7. Closeness coefficients standards and its position.

Trial Number	S+	S-	CC_s	Position
1	0.11046	0.04179	0.27450	23
2	0.09441	0.04827	0.33830	19
3	0.09465	0.05598	0.37165	13
4	0.08026	0.06041	0.42945	9
5	0.07866	0.05230	0.39935	11
6	0.07526	0.06134	0.44905	7
7	0.07415	0.08403	0.53126	4
8	0.05259	0.10503	0.66636	1
9	0.05949	0.09498	0.61486	2
10	0.10027	0.04279	0.29909	22
11	0.10887	0.02737	0.20092	27
12	0.10344	0.04866	0.31994	20
13	0.08596	0.04834	0.35995	15
14	0.09845	0.03358	0.25431	25
15	0.09398	0.04834	0.33966	18
16	0.08114	0.04987	0.38065	12
17	0.07368	0.06213	0.45751	6
18	0.07562	0.05908	0.43862	8
19	0.10253	0.05629	0.35446	16
20	0.09522	0.05393	0.36156	14
21	0.09109	0.06190	0.40462	10
22	0.09843	0.02875	0.22603	26
23	0.09059	0.04050	0.30896	21
24	0.09118	0.04740	0.34204	17
25	0.09375	0.03253	0.25757	24
26	0.06482	0.07595	0.53953	3
27	0.07071	0.06369	0.47389	5

From all the L27 closeness standard values, observation number 8 gave the simplest and the most excellent closeness coefficients with multi-response features. Fig. (**7**) shows the closeness coefficients for every observation.

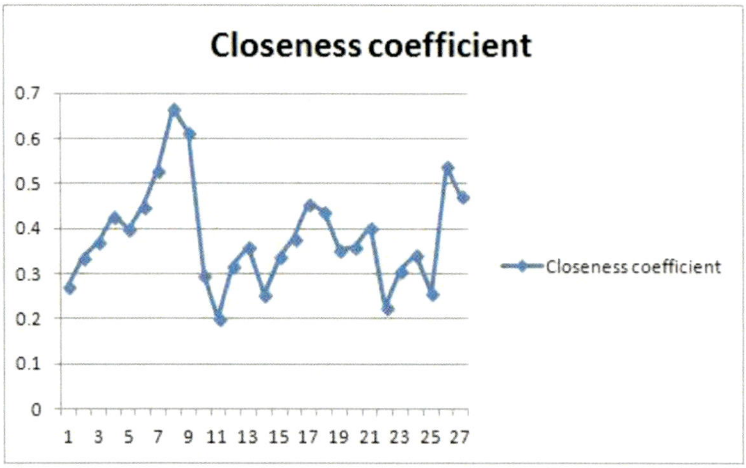

Fig. (7). Closeness coefficient.

A greater coefficient denotes the closeness of the resultant observations to the optimized value. The result showed that observation number 8 provided the most favourable condition for process variables with multi-responses in the CNC turning process.

Table **8** provides process variables with closeness coefficients of the mean response. Mean response values were determined using closeness coefficient values at every observation and level of cutting variables.

Table 8. Closeness coefficient of mean response value.

Machining Variables	Level 1	Level 2	Level 3	Maximum-Minimum Value
cutting spindle speeds or spindle velocity (C_s) (rev/min)	0.33896	0.363184	0.452752	0.113792
Rate of feed (F_r) (mm/min)	0.325002	0.345422	0.484471	0.159469
Depth of cut (D_c) (mm/min)	0.345884	0.391866	0.417146	0.071262
Nose radius(N_r) (mm)	0.358701	0.382618	0.413577	0.054876

From Fig. (**7**), the optimal set of results were obtained based on the closeness coefficient. The most favourable variable was obtained at cutting spindle speeds or spindle velocity (C_s) of 100 rev/min, rate of feed (F_r) of 0.20mm/min (F), depth of cut (D_c) of 1.5mm/min (C), and Nose radius(N_r) of 0.8 mm for an optimal variable setting.

This was achieved by using maximum and minimum values of the closeness coefficient. Feed was the most influencing variable compared to other input variables. The greater the rate of feed, the higher the values of MRR, FX, FZ, and lower FY, and Ra.

ANOVA was used to determine the most significant level of every process variable. ANOVA is an effective way to find the most significant level of every process variable to predict the influencing factors in the CNC turning process. F ratio is the most influencing process variable as determined by a statistical test at each confidence level. The selection of significance level depends on a critical value, and the significant value does not exceed the critical value. ANOVA confirmation test was performed at 95% confidence level and 5% most significant level.

Table **9** shows the results of the ANOVA test, which reveal that the multi-response variable was determined by using input variables. Feed rate and Nose radius were the most influential variables, while other input variables were less influential with output responses like Ra, MRR, FX, FY, and FZ. It was observed that a higher rate of feed produced increased MRR, FX, and FZ and decreased FY and Ra..

Table 9. Results of ANOVA test results.

Source	DF	Adj SS	Adj MS	F-Value	P-Value
Speed	2	4.815	2.4073	9.46	0.002
Feed	2	25.809	12.9044	50.73	0.000
Depth of cut	2	3.848	1.9241	7.56	0.004
Nose radius	2	7.43	3.715	14.6	0.000
Error	18	4.579	0.2544	-	-
Total	26	46.481	-	-	-
s=0.504358	R-sq= 90.15%	R-sq(adj)= 85.77%	R-sq(Pred)= 77.84%	-	-

The most influential variables were found, keeping in mind their order of rate of feed, Nose radius, speed, and depth of cut. Table **9** shows the rate of feed as the most influential variable to affect multi-response characteristics. Table **10** affirms the test results. An Affirmation test was conducted to compare the results between the initial set design and the optimal design of the CNC turning process variables.

Table 10. Affirmation of test results.

Output responses							
Selection variables	Arrangement stage	Ra	MRR	FX	FY	FZ	Cc_s
Initial setting variable	A1B2C3D4	2.37	1.456	187.4	79.77	286.2	0.53126
Optimal conditions							
Predicted	A1B3C3D3						0.60633
Optimal design	A1B3C3D3	2.18	3.723	236.4	77.99	400.4	0.66633

TOPSIS grade improvement: 0.132

After obtaining optimum conditions, a confirmation test was performed for validation which was in turn used to enhance the multi-response characteristics of the CNC turning process. The following equation was used to estimate the predicted value design

$$\gamma_{\text{predicted}} = \gamma_m + \sum_{i=1}^{n}(\gamma_i - \gamma_m) \quad \textbf{(6)}$$

γ_m is the mean value at optimum conditions.

n is the number of inputs.

From the validation results, it was observed that the overall performance of the optimum design condition (A1B3C3D3) had a greater value than that of the Initial design condition (A1B2C3D4) and that the predicted value was close to the optimum design condition.

CONCLUDING REMARKS

In the turning process of AA6061-T6 alloy to obtain multi-response characteristics, optimization was done by the TOPSIS method. Optimum values of the CNC Turning variables were identified by increasing MRR, Fx, Fz, and decreasing Ra, and Fy. Cutting spindle speed or spindle velocity (C_s) of 100 rev/min, rate of feed (F_r) of 0.20 mm/min, depth of cut(D_c) of 1.0 mm/min, and Nose radius(N_r) of 0.4 mm at observation number 8 showed a higher closeness coefficient value. Cutting spindle speed or spindle velocity (C_s) of 100 rev/min, rate of feed (F_r) of 0.20mm/min (F), depth of cut(D_c) of 1.5mm/min (C), and Nose radius(N_r) of 0.8 mm of the process variables produced an optimum value. ANOVA was used to find the most favourable process variable for the CNC's turning of multi-performance characteristics.

It was inferred from the observations that the rate of feed was the highest influencing variable preceded by Nose radius, speed, and depth of cut. Both TOPSIS and ANOVA were found to be effective tools for tuning multi-response problems than previously used methods.

CONSENT FOR PUBLICATION

Not applicable.

CONFLICT OF INTEREST

The author declares no conflict of interest, financial or otherwise.

ACKNOWLEDGEMENTS

The authors are thankful to Mepco Schlenk Engineering College, Sivakasi and Sri Vidya College of Engineering and Technology, virudhunagar, K.Ramakrishnan College of Engineering, Trichy, for providing research facilities to carry out this research work.

REFERENCES

[1] S. Ramesh, R. Viswanathan, and S. Ambika, "Measurement and optimization of surface roughness and tool wear via grey relational analysis, TOPSIS and RSA techniques", *Measurement,* vol. 78, pp. 63-72, 2016.
[http://dx.doi.org/10.1016/j.measurement.2015.09.036]

[2] A. Khan, and K. Maity, "Application of MCDM-based TOPSIS method for the optimization of multi quality characteristics of modern manufacturing processes", *Int. J. Eng. Res. Africa.,* vol. 23, pp. 33-51, 2016.
[http://dx.doi.org/10.4028/www.scientific.net/JERA.23.33]

[3] B. Wang, Z. Liu, Q. Song, Y. Wan, and Z. Shi, "Proper selection of cutting parameters and cutting tool angle to lower the specific cutting energy during high speed machining of 7050-T7451 aluminum alloy", *J. Clean. Prod.,* vol. 129, pp. 292-304, 2016.
[http://dx.doi.org/10.1016/j.jclepro.2016.04.071]

[4] Soorya Prakash, K., Gopal P. M., and Karthik S., "Multi-objective optimization using Taguchi based grey relational analysis in turning of Rock dust reinforced Aluminum MMC", *Measurement,* vol. 157, p. 107664, 2020.

[5] M. Dobrzynski, D. Chuchala, and K.A. Orlowski, "The effect of alternative cutter paths on flatness deviations in the face milling of aluminum plate parts", *J. Mach. Eng.,* vol. 18, no. 1, pp. 80-87, 2018.
[http://dx.doi.org/10.5604/01.3001.0010.8825]

[6] H.A. Rahman, M.H. Fauziah, M.N. Rizal, O. Muhamad, M.M. Mahadzir, S.O. Mohamed, M.S.M. Zameri, and S. Safian, "Investigation of energy consumption during milling operation", In: *In AIP Conference Proceedings* vol. 2129. , 2019, no. 1, p. 020009. AIP Publishing LLC.
[http://dx.doi.org/10.1063/1.5118017]

[7] Z. Sun, S. To, S. Zhang, and G. Zhang, "Theoretical and experimental investigation into non-uniformity of surface generation in micro-milling", *Int. J. Mech. Sci.,* vol. 140, pp. 313-324, 2018. [http://dx.doi.org/10.1016/j.ijmecsci.2018.03.019]

[8] S. Wojciechowski, and K. Mrozek, "Mechanical and technological aspects of micro ball end milling with various tool inclinations", *Int. J. Mech. Sci.,* vol. 134, pp. 424-435, 2017. [http://dx.doi.org/10.1016/j.ijmecsci.2017.10.032]

[9] S. Wojciechowski, M. Wiackiewicz, and G.M. Krolczyk, "Study on metrological relations between instant tool displacements and surface roughness during precise ball end milling", *Measurement,* vol. 129, pp. 686-694, 2018. [http://dx.doi.org/10.1016/j.measurement.2018.07.058]

[10] R.L. malghan, K.M.C. Rao, S. ArunKumar, S.S. Rao, and R.J. D'Souza, "R.L. malghan, K.M.C. Rao, S. ArunKumar, S.S. Rao, R.J. D'Souza, Effect of process parameters in face milling operation and analysis of cutting force using indirect method", *Mater. Manuf. Process.,* vol. 33, no. 13, pp. 1406-1414, 2018. [http://dx.doi.org/10.1080/10426914.2017.1388520]

[11] R.B.D. Pereira, R.R. Leite, A.C. Alvim, A.P. de Paiva, J.R. Ferreira, and J.P. Davim, "Multi-objective robust optimization of the sustainable helical milling process of the aluminum alloy Al 7075 using the augmented-enhanced normalized normal constraint method", *J. Clean. Prod.,* vol. 152, pp. 474-496, 2017. [http://dx.doi.org/10.1016/j.jclepro.2017.03.121]

[12] S. Nipanikar, V. Sargade, and R. Guttedar, "Optimization of process parameters through GRA, TOPSIS and RSA models", *Int. J. Ind. Eng. Comput.,* vol. 9, pp. 137-154, 2018. [http://dx.doi.org/10.5267/j.ijiec.2017.3.007]

[13] A. Thakur, A. Manna, and S. Samir, *Multi-Response Optimization of Turning Parameters during Machining of EN-24 Steel with SiC Nanofluids Based Minimum Quantity Lubrication.* Silicon, 2020. [http://dx.doi.org/10.1007/s12633-019-00102-y]

[14] M.A. Xavior, M. Patil, A. Maiti, M. Raj, and N. Lohia, "Machinability studies on INCONEL 718", *In IOP conference series: materials science and Engineering,* vol. 149, 2016no. 1, p. 012019 Publishing, 2016. IOP. [http://dx.doi.org/10.1088/1757-899X/149/1/012019]

[15] R.F. Garcia, E.C. Feix, H.T. Mendel, A.R. Gonzalez, and A.J. Souza, "Optimization of cutting parameters for finish turning of 6082-T6 aluminum alloy under dry and RQL conditions", *J. Braz. Soc. Mech. Sci. Eng.,* vol. 41, no. 8, p. 317, 2019. [http://dx.doi.org/10.1007/s40430-019-1826-4]

[16] M. Jebaraj, and M. Pradeep Kumar, "Effect of cryogenic CO_2 and LN_2 coolants in milling of aluminum alloy"., *Mater. Manuf. Process.,* vol. 34, no. 5, pp. 511-520, 2019. [http://dx.doi.org/10.1080/10426914.2018.1532591]

[17] K. Xhanari, M. Finšgar, M. Knez Hrnčič, U. Maver, Ž. Knez, and B. Seiti, "Green corrosion inhibitors for aluminium and its alloys: A review", *RSC Advances,* vol. 7, no. 44, pp. 27299-27330, 2017. [http://dx.doi.org/10.1039/C7RA03944A]

[18] T. Paramjit, J. A. Khan, M. Dongre, and S. Gupta, "Multi Objective Optimization in CNC End Milling of Inconel 718 Super Alloy by Taguchi-Grey-Fuzzy Method", *In 2019 International Conference on Nascent Technologies in Engineering (ICNTE),* pp. 1-4, 2019. IEEE. [http://dx.doi.org/10.1109/ICNTE44896.2019.8945898]

[19] N. Yuvaraj, and M.P. Kumar, "Investigation of process parameters influence in abrasive water jet cutting of D2 steel", *Mater. Manuf. Process.,* vol. 32, no. 2, pp. 151-161, 2017. [http://dx.doi.org/10.1080/10426914.2016.1176183]

[20] A. Shokrani, V. Dhokia, and S.T. Newman, "Energy conscious cryogenic machining of Ti-6Al-4V

titanium alloy", *Proc. Inst. Mech. Eng., B J. Eng. Manuf.,* vol. 232, no. 10, pp. 1690-1706, 2018. [http://dx.doi.org/10.1177/0954405416668923]

[21] G. Karthik Pandiyan, and T. Prabaharan, "Optimization of machining parameters on AA6351 alloy steel using Response Surface Methodology (RSM)", *Mater. Today Proc.,* vol. 1–4, pp. 2686-2689, 2020. [http://dx.doi.org/10.1016/j.matpr.2020.01.369]

[22] G. Karthik Pandiyan, and T. Prabaharan, "Soft Computing Method For Prediction of Surface Roughness inMachining Parameters in Turning Process of Aluminium Metal Matrix Composites", *Interciencia,* vol. 45, pp. 69-80, 2020.

CHAPTER 2

Machinability Analysis and Optimization by the Taguchi-gra-pca Approach of PMEDM Process Parameters of Monel K 500

Jafrey Daniel James D.[1,*], **N. Ganesh**[1] and **Pandian G. Karthik**[2]

[1] *Department of Mechanical Engineering, K Ramakrishnan College of Engineering, Samayapuram, Trichy, Tamilnadu 621112, India*

[2] *Department of Mechanical Engineering, Sri Vidya College of Engineering and Technology, Virudhunagar 626005, Tamilnadu, India*

Abstract: Powder mixed electrical discharge machining (PMEDM) is applied for machining complex parts that cannot be machined by conventional machining processes. In this work, the optimization of PMEDM process parameters for Monel K 500 was performed, and its impact was analysed. PMEDM experiments were planned using the L'9 orthogonal array with pulse-on-time, pulse-off-time, and current as input parameters. The output parameters were material removal rate (MRR), tool wear rate (TWR), and surface roughness. Using a combination of principal component analysis (PCA) and grey relational analysis (GRA), the outcome was optimized. The analysis of variance was used to analyze the impact of input parameters over the output parameters. Finally, confirmation tests were performed to confirm the optimized output values of the machining process.

Keywords: Analysis of Variance, Current, Eigen Values, Eigen Vectors, Electric Discharge, Grey Relational Analysis, Machining, Machining Time, Monel K 500, Principal Component Analysis, Pulse-Machining Process, Pulse-off-current, Pulse-on-current, Signal-to-Noise Ratio, Significant, Surface Roughness, Taguchi, Tool Wear Rate, Voltage.

INTRODUCTION

Metal matrix composites find applications in several industries such as the automobile sector, structural units, and various other fields. It is more common that metal matrix composites are machined before utilization for end applications. The most widely used non-traditional machining is electric discharge machining (EDM).

[*] **Corresponding author Jafrey Daniel James D:** Department of Mechanical Engineering, K Ramakrishnan College of Engineering, Samayapuram, Trichy, Tamilnadu 621112, India; E-mail:jafrey.daniel@gmail.com

C. Samson Jerold Samuel, M. Suresh & S. Gnanasekaran (Eds.)

EDM is used in a number of sectors such as aerospace, biomedicine, and nuclear fields [1]. Occasionally, problems of poor surface finish and low machining efficiency arise, restricting its usage in several applications. To overcome this, a powder is introduced to the dielectric medium, which is termed as PMEDM [2]. The powder added to the dielectric medium increases the distance between the workpiece and electrode. The formation of excess thermal energy from the sparks results in heating, which causes the material to melt and vaporize from the workpiece's surface [3].

During PMEDM, an electric field energizes the spark gap, which is filled with powder particles suspended in a dielectric fluid. The electric field accelerates the charged particles, enabling them to act as conductors. Powder particles aggregate and settle as clusters under the sparking layer [4, 5]. Boron carbide powders were used by Kolli and Kumar, along with the dielectric medium, on machine titanium alloys [6]. They concluded that low surface roughness values were produced by adding 15 grams of boron carbide powder. When the aluminum powder was added in a range of 300 to 400 μm to the dielectric medium, there was an increase in MRR values and a decrease in TWR and SR. The addition of SiC powder during EDM of tungsten carbide increased productivity [7].

Conventional optimization processes generally use one parameter while keeping the remaining parameters stable, thus making the process complex. Using statistical experiment design is the best way to optimize multiple variables [8], and some of the commonly used methods for experimental design include response surface methodology and Taguchi. Among the above-mentioned techniques, Taguchi is commonly used in a limited number of experiments for optimization [9]. It reduces cost and the number of experiments, thus achieving output with a lower number of trials. Optimization conditions of metal-contaminated water using Taguchi methodology were reported by Barrado *et al.* [10]. Moreover, many researchers used the Taguchi method to plan and analyze experiments [11].

Deng implemented a grey method in 1982 to solve complex problems with fewer details. GRA was used in the past decade to solve multiple-characteristic problems [12]. Pan *et al.* [13] used Taguchi along with GRA to refine the cutting parameters for the welding of titanium sheets. Performance features of EDM were optimized using GRA [14]. It was inferred from their findings that GRA could be used effectively to solve multiple properties. By merging GRA and cuckoo search algorithm, Kumar and Panneerselvam [15] optimized the wear characteristics of nylon with 30 percent glass fiber content.

Pearson first presented PCA in 1901, and Hotelling reframed it as a statistical method in 1993. This simplified multiple associated variables into less uncorrelated and separate elements, retaining all original information as much as possible by using a linear combination. Presently, PCA is also used to solve multi-objective problems. Coupled with GRA, HS *et al.* [16] used PCA to refine cutting parameters of the end milling process. They concluded that manufacturers could utilize the results of the PCA process. Together with GRA and PCA, Kaushik and Singhal [17] used Taguchi to refine the wear parameters of composites based on aluminium.

From an extensive literature survey, it was found that there are no works related to PMEDM of Monel K 500. The present research work focuses on the analysis and optimization of process parameters for the PMEDM of Monel K 500. Input parameters considered for the experiment were (i) pulse-on-time (μs), (ii) pulse-off-time (μs) and (iii) current (A). Output characteristics studied were (i) MRR, (ii) surface roughness, and (iii) TWR. Taguchi was used for experimental design, while GRA and PCA were used to optimize process parameters and find the optimized values.

MATERIALS AND METHODS

Monel K 500, which has applications in the manufacturing of marine fixtures, pumps, and valves, was selected as the material for the PMEDM process. Properties of the Monel K 500 material are listed in Table **1**. A silver-coated copper tool was selected for PMEDM. Kerosene was used as the electrolyte, and the powder added was nano aluminium. The powder was mixed with a dielectric fluid using a magnetic stirrer, which completely mixed the powder. The pump was equipped with an EDM system that ensured powder particles did not settle in the tank. The experiments were planned in accordance with the L'9 orthogonal array. Input parameters considered for PMEDM were (i) pulse-on-time (2, 3, 4 (μs)), (ii) pulse-off-time (8, 9, 10 (μs)), and (iii) current (2, 3, 4 (A)). The parameters measured as output were (i) MRR, (ii) surface roughness, and (iii) TWR. MRR was evaluated using Equation (1). A surface roughness tester (TR 200, Time Group Inc.) with a display resolution of 0.001 μ was used to measure the surface roughness of the machined surface. TWR for the tool was measured using Equation (2). Finally, the measured output responses were optimized using a combination of GRA and PCA. Confirmation experiments were performed to validate the optimized values.

$$MRR = \frac{Weight\ of\ WP\ before\ Machining - Weight\ of\ WP\ after\ Machining}{Machining\ Time} \quad \textbf{(1)}$$

$$TWR = \frac{Initial\ tool\ wear\ rate - Final\ tool\ wear\ rate}{Machining\ Time} \quad \textbf{(2)}$$

Table 1. Monel K 500 chemical composition (wt. %).

Material	Chemical Composition (%)
Nickel	63.0 Min
Carbon	0.25 Max
Manganese	1.5 Max
Iron	2.0 Max
Silicon	0.5 Max
Copper	27.0 - 33.0
Aluminium	2.30 – 3.15

RESULTS AND DISCUSSIONS

PMEDM experiments were carried out according to the L'9 orthogonal array, and their values are listed in Table **2**. Output parameters were measured, and their optimization processes are discussed in the following sections.

Table 2. Input and Output Process parameters.

Exp. No	Input Process Factors			Output Characteristics		
	Pulse-on-time (μs)	Pulse-off-time (μs)	Current (A)	MRR (g/cm^3)	Ra (μm)	TWR (g/cm^3)
1	1500	0.1	0.1	0.0101	2.5	0.13
2	1500	0.15	0.2	0.00954	1.065	0.21
3	1500	0.2	0.3	0.009	2.25	0.2
4	1750	0.1	0.2	0.0112	2.114	0.11
5	1750	0.15	0.3	0.0113	2.09	0.15
6	1750	0.2	0.1	0.0099	1.989	0.11
7	2000	0.1	0.3	0.011	3.196	0.14
8	2000	0.15	0.1	0.0099	2.527	0.16
9	2000	0.2	0.2	0.01	1.406	0.15

Material Removal Rate

MRR is one of the vital factors that decide the finishing of a machined product. In the current experiment, the MRR values were very minimum, as given in Table **2**.

The main effect plots of MRR with respect to pulse-on-time, pulse-off-time, and current are depicted in Figure 1, which reveals that MRR increased up to 2μs with an increase in pulse-on-time and then decreased with an increase of pulse-on-time. When pulse-on-time increased, heat flux was applied for a longer period of time. This phenomenon increased heat in the work piece due to plasma channel expansion, causing an increase in MRR. Pressure within the plasma channel decreased as pulse-on-time increased further, resulting in a reduction of MRR values. Hence, MRR values were minimum when pulse-on-time was maximum. Pulse-off-time is defined as the time during which the dielectric is deionized. Pulse-off-time main effect plots with respect to MRR are shown in Fig. (**1**). Maximum MRR value was reached when pulse-off-time was at a lower level. When the pulse-off time was less, there occurred a short circuit and an arcing process. Once the pulse-off-time decreased, the gel was reduced. Fig. (**1**) shows that when the peak current value increased, MRR value also increased. When peak current was set at 4A, the diameter of the pulse discharge energy channel increased, which in turn increased the diameter and depth of the crater, thus promoting a rise in MRR. Similar findings were also published by Wong *et al.* [18].

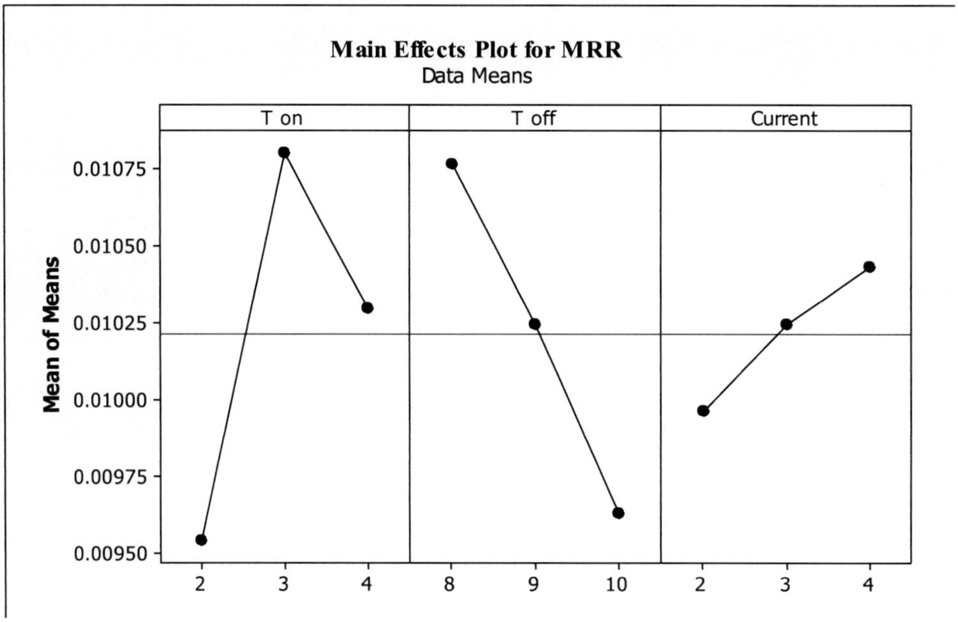

Fig. (1). Main effect plots of MRR.

Tool Wear Rate

Main effect plots for TWR are shown in Fig. (**2**) with respect to pulse-on-time, pulse-off-time and current and their values in Table **2**. We infer from Fig. (**2**) that TWR values decreased with an increase in pulse-on-time. When pulse-on-time was less, the number of negatively charged particles in motion was less and the positive electrode of the device strikes, thus increasing the electrode's fusion rate. Reduction in TWR values at high pulse-on-time was due to a longer pulse-on-time, which increased energy discharge and elevated discharge frequencies resulted in a more conductive energy loss. Main effect plots for pulse-off-time on TWR are shown in Fig. (**2**), from which we infer that TWR increased with a decrease in pulse-off-time values. When pulse-off-time was at the lowest level, the arcing probability was higher at the beginning resulting in the removal of additional material. Material removed could not be flushed away properly so that the gap between the electrode and the workpiece remained stable, resulting in a decrease in TWR at a higher pulse-off-time. Fig. (**2**) illustrates an increase in TWR with an increase in current. When current was set at 2A, less heat energy was produced, resulting in less TWR. Heat generation was higher when the current was set at 4A, resulting in more TWRs. Similar kinds of results had also been reported by Aghdeab [19].

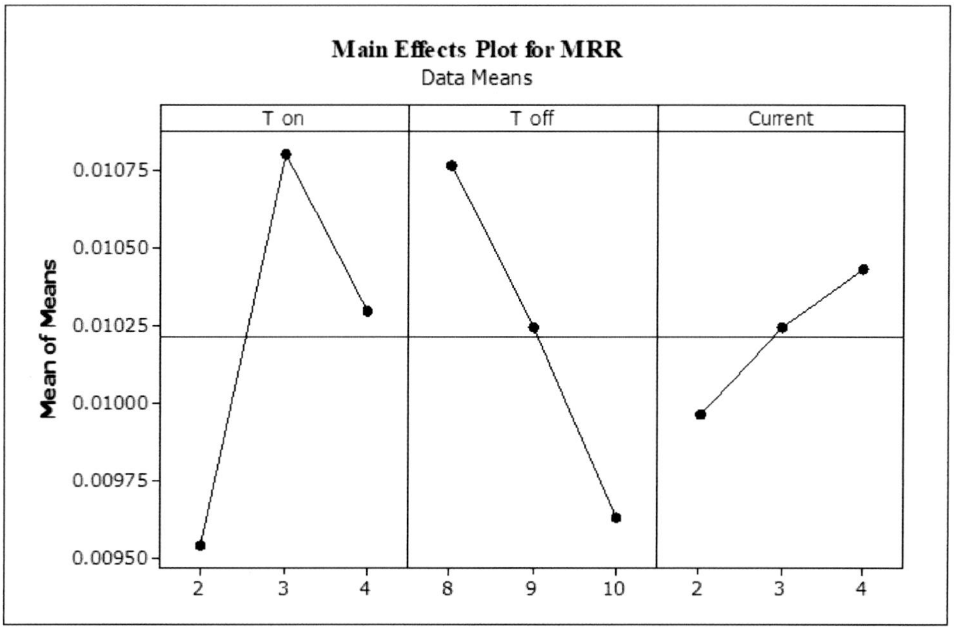

Fig. (2). Main effect plots of TWR.

Surface Roughness

Main effects plots for surface roughness are shown in Fig. (**3**). The MRR mechanism in EDM was due to the formation of sparks and craters, which led to the formation of a rough surface. Surface quality was determined based on energy sparks and crater size. It can be seen that when pulse-on-time increased, MRR was high, which resulted in high surface roughness. It can be inferred from Fig. (**3**) that the surface roughness value was maximum at the lowest pulse-off-time and minimum at a higher pulse-off-time. Fig. (**3**) shows that surface roughness values decreased when the pulse-on-current increased from 2 to 3A. As pulse-on-current proceeded further from 3 to 4A, a molten metal pool was created due to overheating. The surface-formed molten metal evaporated to produce gas bubbles that erupted when the discharge occurred. The material explosion resulted in the removal of the fused material from the surface; craters and pocket marks were produced when the discharge was repeated. This led to an increase in surface roughness. The obtained results were very similar to those obtained by Palanisamy *et al.* [20].

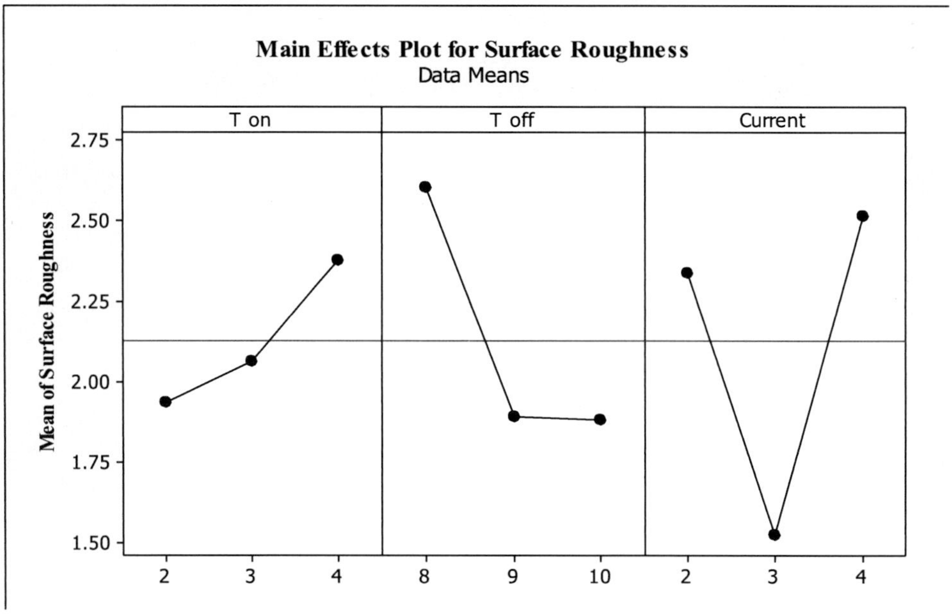

Fig. (3). Main effect plots of Surface Finish.

Grey Relational Analysis

The steps followed during the GRA process were [21].

1. Experiment values were collected and tabulated, as shown in Table **2**
2. Raw output values data was converted to the S/N ratio, as shown in Table **3**.
3. Normalization of data was carried out. Usually, normalization has three categories such as 'smaller-the-better, higher-the-better, and nominal-the-better. In the present work, smaller-the-better was used for surface roughness and machining time. Higher-the-better was applied for MRR. They were calculated using equations (3) and (4).

Table 3. Raw data and S/N ratio of different performance characteristics.

Exp. No	Input Process Factors			S/N ratios		
	MRR (g/cm^3)	Ra (μm)	TWR (g/cm^3)	MRR	Ra	TWR
1	0.0101	2.5	0.13	-39.9136	-7.9588	17.7211
2	0.00954	1.065	0.21	-40.4090	-0.5470	13.5556
3	0.009	2.25	0.2	-40.9151	-7.0437	13.9794
4	0.0112	2.114	0.11	-39.0156	-6.5021	19.1721
5	0.0113	2.09	0.15	-38.9384	-6.4029	16.4782
6	0.0099	1.989	0.11	-40.0873	-5.9727	19.1721
7	0.011	3.196	0.14	-39.1721	-10.0921	17.0774
8	0.0099	2.527	0.16	-40.0873	-8.0521	15.9176
9	0.01	1.406	0.15	-40.0000	-2.9597	16.4782

For higher-the-better,

$$x_i^*(k) = \frac{max\, x_i^0 - x_i^0(k)}{\max\, x_i^0(k) - \min x_1^0(k)} \quad \textbf{(3)}$$

For lower-the-better,

$$x_i^*(k) = \frac{x_i^0(k) - \min x_i^0(k)}{\max x_i^0(k) - \min x_i^0(k)} \quad \textbf{(4)}$$

It can also be done by taking into account the original sequence value and dividing it by the first value in the given order, as shown in equation 5.

$$x_i^*(k) = \frac{x_i^*(k)}{x_i^0(1)} \quad \textbf{(5)}$$

The normalization and sequential normalization are shown in Table **4**.

Table 4. Normalized data of response characteristics and corresponding Sequence Normalization.

Exp. No	Normalization			Sequential Normalization		
	MRR	Ra	TWR	MRR	Ra	TWR
1	0.50669	0.7765	-0.7417	0.49331	0.2235	1.74165
2	0.25604	0	0	0.74396	1	1
3	0	0.68062	-0.0755	1	0.31938	1.07545
4	0.96094	0.62389	-1	0.03906	0.37611	2
5	1	0.6135	-0.5203	0	0.3865	1.52035
6	0.4188	0.56843	-1	0.5812	0.43157	2
7	0.88177	1	-0.627	0.11823	0	1.62705
8	0.4188	0.78628	-0.4205	0.5812	0.21372	1.42054
9	0.46296	0.25277	-0.5203	0.53704	0.74723	1.52035

4. Grey relational coefficient was calculated using Equations (6-9) as shown in Table **5**.

$$\xi(k) = \frac{\Delta min + \xi \Delta max}{\Delta oi(k) + \xi \Delta max} \quad \textbf{(6)}$$

$$\Delta oi = \| x_o^*(k) - x_i^0(k) \| \quad \textbf{(7)}$$

$$\Delta \min = min_{\forall j \varepsilon i} \, min_{\forall k} \, \| x_o^*(k) - x_j^0(k) \| \quad \textbf{(8)}$$

$$\Delta \max = max_{\forall j \varepsilon i} \, max_{\forall k} \, \| x_o^*(k) - x_j^0(k) \| \quad \textbf{(9)}$$

where Δoi- the absolute difference between $x_i^0(k)$ and $x_i^*(k)$ and it can be called the sequence of deviation. The term ξ represents the distinguishing coefficient. Mostly, ξ value is selected as 0.5.

Table 5. Grey Relational Coefficient of response characteristics.

Exp. No	Grey Relational Coefficient		
	MRR	Ra	TWR
1	0.50337	0.69109	0.22305
2	0.40194	0.33333	0.33333
3	0.33333	0.61022	0.31737
4	0.92754	0.5707	0.2
5	1	0.56402	0.24748
6	0.46245	0.53673	0.2
7	0.80876	1	0.23507
8	0.46245	0.70055	0.26034
9	0.48214	0.40089	0.24748

Principal Component Analysis

Hotelling and Pearson designed PCA, which discusses the covariance and variance of all performance features [16]. The steps related to PCA are listed below,

(i) For the construction of covariance and variance, normalized values were used, as shown below in Equation 10. In the above equation, m is the number of experimental tests, and n is the number of responses. The number of experimental runs in the present analysis was 9, and the number of output responses 3.

(ii) $X_i(j), i = 1\ to\ m, j = 1\ to\ n$

$$X_i = \begin{matrix} X_1(1) & X_1(2) \dots\dots\dots\dots\dots & X_1(n) \\ X_2(1) & X_2(2) \dots\dots\dots\dots\dots & X_2(n) \\ \cdots\cdots & \cdots\cdots & \cdots\cdots \\ \cdots\cdots & \cdots\cdots & \cdots\cdots \\ X_m(1) & X_m(2) \dots\dots\dots\dots\dots & X_m(n) \end{matrix} \quad (10)$$

In the above equation, m is the number of experimental tests, and n is the number of responses. The number of experimental runs in the present analysis is 9, and the number of output responses is 3.

2. Correlation coefficient was calculated using Equation 11 below.

$$R_{jl} = \left(\frac{cov(X_j(j),\ X_i(l))}{\sigma_{Xi}(j) X \sigma_{Xi}(l)}\right) \quad \textbf{(11)}$$

where j = 1 to n and l = 1 to n.

In the above equation $cov(X_j(j), X_i(l))$ are the sequences of the covariance $X_i(j)$ and $X_i(l)$, $\sigma_{Xi}(j)$ is the S.D of the sequences $X_i(j)$ and $\sigma_{Xi}(j)$ represents the S.D of the sequence $X_i(l)$3. Eigen values and Eigen vectors were calculated using a correlation coefficient array as shown in Equation 12. Eigen values and the explained variation are shown in Table **6**, from which it can be inferred that the first principal component characterizing all output characteristics was 69.2%.

Table 6. Eigen values and explained variation.

Principal Component	Eigen Value	Explained Variation (%)
PC1	2.0750	69.2
PC2	0.7599	25.3
PC2	0.1651	5.55

$$(R - \lambda_k I_m)V_{ik} = 0 \quad \textbf{(12)}$$

4. In the above equation $\sum_{k=1}^{n} \lambda_k = n$ where k ranges from 1 to n and $V_{ik} = [a_{k1}\ a_{k1} \ldots\ldots\ldots a_{kn}]\ T$ are the eigen vectors corresponding to eigen values λ_K The below Equation 13 was used for the calculation of principal components.

$$y_{mk} = \sum_i^n X_m(i) * V_{ik} \quad \textbf{(13)}$$

In the above equation, the first principal component is denoted asy_m1 and the second by y_m2. Eigen vectors for the principal components are shown in Table **7**. To find the relative significance of all output performance characteristics, GRA values were given as input for PCA. For each output value, weighted values were determined using PCA. Table 7 shows the performance attributes of each GRA. These values were used to assess the correlation coefficient matrix and their Eigen values by applying equations listed in Table **7**. Table **7** shows the weight of each output obtained for the corresponding vector values of the principal component. GRG was determined after multiplying PCA values and their weight performance characteristics, as shown in Table **8**.

Table 7. Eigenvectors for principal components.

Performance Characteristics	Eigen Vector		
	PC1	PC2	PC3
Surface Roughness	0.503	-0.769	-0.395
Tool Wear Rate	0.661	0.047	0.749
Material Removal Rate	0.557	0.638	-0.532

It can be seen from Table **8** that GRG values were within a range from 0 and 1. There was a relationship between comparability and reference sequence as GRG values moved toward the maximum.

Table 8. Grey Relational Grade and its Rank.

Exp. No	Grey Relational Grade	Rank
1	0.147811	5
2	0.119353	9
3	0.13247	6
4	0.185915	3
5	0.200166	2
6	0.126854	7
7	0.212394	1
8	0.148433	4
9	0.123796	8

Table 9. Main effect plots of GRG.

Level	Pulse-on-time	Pulse-off-time	Current
1	0.1332	0.1820*	0.1410
2	0.1710*	0.1560	0.1430
3	0.1615	0.1277	0.1817*
Delta	0.0378	0.0543	0.0406
Rank	3	1	2

Confirmatory Experiment

As the obtained optimum set of parameters did not fall on any set of experiments, a confirmation experiment was conducted to validate it. The results of the confirmation experiment were compared with the initial experiments, as shown in Table **10**. Table **10** shows an improvement in the mean values and S/N ratio of the optimized one when compared to the initial experiments. As a result, these confirmation tests showed that the proposed method was good to solve multiple characteristics of PMEDM for Monel K 500. In the present work, MRR improved with a decrease in TWR and surface roughness.

Table 10. Comparison between initial level and optimal level.

Factors	Combination	Mean Values	S/N ratio
Initial Design	Pulse-on-time 2 (μs) Pulse-off-time -8 (μs) Current – 2 (A)	0.1457	-16.733
Optimal Design	Pulse-on-time -2 (μs) Pulse-off-time -8 (μs) Current – 4 (A)	0.1864	-14.692

Analysis of Variance

Analysis of Variance (ANOVA) is usually applied to study the significance of factors [22] and [23]. ANOVA values of GRG are shown in Table **11**. All parameters were statistically significant as the calculated F-value was more than the F-table value. From Table **11**, it can be inferred that pulse-off-time was the dominant factor followed by current and pulse-on-time. Error percentage during ANOVA determined whether the experiment was feasible or not. It depended on uncontrollable conditions. The error must be less than 15% to make the experiment acceptable. Error percentage in the current study was 1.57%, proving that the experiment was valid. The findings observed were also consistent with the results suggested by HS *et al.* [16]

Table 11. ANOVA for GRG.

Source	DOF	Adj. SS	Adj. MS	F_{cal}	% of Contribution
Pulse-on-time	**2**	0.0023180	0.0011590	78.55	23.34
Pulse-off-time	2	0.0044306	0.0022153	150.14	43.37
Current	2	0.0031501	0.0031501	106.75	31.72
Error	2	0.0000295	0.0000295	-	1.57
Total	8	0.0099282	-	-	-

CONCLUSION

In the current work, machining of Monel K 500 was carried out using PMEDM. Experiments were designed using Taguchi techniques, and output was optimized using GRA and PCA approaches. The results are as follows:

(i) Machining of Monel K 500 using PMEDM with the suspension of nano-aluminum powder in a dielectric fluid resulted in improved MRR and a reduction in machining time and surface roughness.

(ii) Output of the PMEDM process was optimized using a combination of GRA and PCA approaches. The optimized set of process parameters was pulse-on-current at 3 μs, pulse-of-current at 8 μs, and current at 4A. A confirmation experiment revealed an improvement in the values.

(iii) GRG values of ANOVA suggested pulse-off-time as the dominant factor at 43.37%, followed by the current at 31.72% and pulse-on-time at 23.37%.

(iv) Further work can be carried out by using powder of lesser size and observing machining features.

CONSENT FOR PUBLICATION

Not applicable.

CONFLICT OF INTEREST

The author declares no conflict of interest, financial or otherwise.

ACKNOWLEDGEMENTS

Declared none.

REFERENCES

[1] F. Müller, and J. Monaghan, "Non-conventional machining of particle reinforced metal matrix 2. composite", *Int. J. Mach. Tools Manuf.*, vol. 40, no. 9, pp. 1351-1366, 2000. [http://dx.doi.org/10.1016/S0890-6955(99)00121-2]

[2] H.M. Chow, B.H. Yan, F.Y. Huang, and J.C. Hung, "Study of added powder in kerosene for the micro-slit machining of titanium alloy using electro-discharge machining", *J. Mater. Process. Technol.*, vol. 101, no. 1-3, pp. 95-103, 2000. [http://dx.doi.org/10.1016/S0924-0136(99)00458-6]

[3] A. Satpathy, S. Tripathy, N.P. Senapati, and M.K. Brahma, "Optimization of EDM process parameters for AlSiC-20% SiC reinforced metal matrix composite with multi response using TOPSIS", *Mater. Today Proc.*, vol. 4, no. 2, pp. 3043-3052, 2017. [http://dx.doi.org/10.1016/j.matpr.2017.02.187]

[4] S. Mohanty, A. Mishra, B.K. Nanda, and B.C. Routara, "Multi-objective parametric optimization of

nano powder mixed electrical discharge machining of AlSiCp using response surface methodology and particle swarm optimization", *Alex. Eng. J.,* vol. 57, no. 2, pp. 609-619, 2018.
[http://dx.doi.org/10.1016/j.aej.2017.02.006]

[5] S. Singh, J. Kumar, and S. Kumar, "Influences of process parameters on MRR improvement in simple and powder-mixed EDM of AA6061/10% SiC composite", *Mater. Manuf. Process.,* vol. 30, no. 3, pp. 303-312, 2015.
[http://dx.doi.org/10.1080/10426914.2014.930888]

[6] M. Kolli, and A. Kumar, "Effect of boron carbide powder mixed into dielectric fluid on electrical discharge machining of titanium alloy", *Proc. Med. Sci.,,* vol. 5, pp. 1957-1965, 2014.
[http://dx.doi.org/10.1016/j.mspro.2014.07.528]

[7] B.T. Long, N.H. Phan, N. Cuong, and V.S. Jatti, "Optimization of PMEDM process parameter for maximizing material removal rate by Taguchi's method", *Int. J. Adv. Manuf. Technol.,* vol. 87, no. 5-8, pp. 1929-1939, 2016.
[http://dx.doi.org/10.1007/s00170-016-8586-4]

[8] D.J. Daniel, and K. Panneerselvam, "Abrasive wear of polypropylene/Cloisite 30B/Elvaloy AC 3427 nanocomposites", *J. Compos. Mater.,* vol. 52, no. 13, pp. 1833-1843, 2018.
[http://dx.doi.org/10.1177/0021998317734624]

[9] R. Pundir, G.H.V.C. Chary, and M.G. Dastidar, "Application of Taguchi method for optimizing the process parameters for the removal of copper and nickel by growing Aspergillus sp", *Water Resour. Ind.,* vol. 20, pp. 83-92, 2018.
[http://dx.doi.org/10.1016/j.wri.2016.05.001]

[10] E. Barrado, M. Vega, R. Pardo, P. Grande, and J.L. Del Valle, "Optimisation of a purification method for metal-containing wastewater by use of a Taguchi experimental design", *Water Res.,* vol. 30, no. 10, pp. 2309-2314, 1996.
[http://dx.doi.org/10.1016/0043-1354(96)00119-4]

[11] H. Ramakrishnan, R. Balasundaram, P. Selvaganapathy, M. Santhakumari, P. Sivasankaran, and P. Vignesh, "Experimental investigation of turning Al 7075 using Al_2O_3 nano-cutting fluid: ANOVA and TOPSIS approach"., *SN Applied Sciences,* vol. 1, no. 12, p. 1639, 2019.
[http://dx.doi.org/10.1007/s42452-019-1664-0]

[12] N. Manikandan, R. Raju, D. Palanisamy, and J.S. Binoj, "Optimisation of spark erosion machining process parameters using hybrid grey relational analysis and artificial neural network model", *Int. J. Mach. Mach. Mater.,,* vol. 22, no. 1, pp. 1-23, 2020.
[http://dx.doi.org/10.1504/IJMMM.2020.104007]

[13] L.K. Pan, C.C. Wang, S.L. Wei, and H.F. Sher, "Optimizing multiple quality characteristics via Taguchi method-based Grey analysis", *J. Mater. Process. Technol.,* vol. 182, no. 1-3, pp. 107-116, 2007.
[http://dx.doi.org/10.1016/j.jmatprotec.2006.07.015]

[14] J.L. Lin, and C.L. Lin, "The use of the orthogonal array with grey relational analysis to optimize the electrical discharge machining process with multiple performance characteristics", *Int. J. Mach. Tools Manuf.,* vol. 42, no. 2, pp. 237-244, 2002.
[http://dx.doi.org/10.1016/S0890-6955(01)00107-9]

[15] S. Kumar, and K. Panneerselvam, "Research on tribological behaviors of pure and glass fiber reinforced nylon 6 composites against polymer disc", *Journal of Material Science and Mechanical Engineering,* vol. 2, no. 6, pp. 24-28, 2015.

[16] H.S. Lu, C.K. Chang, N.C. Hwang, and C.T. Chung, "Grey relational analysis coupled with principal component analysis for optimization design of the cutting parameters in high-speed end milling", *J. Mater. Process. Technol.,* vol. 209, no. 8, pp. 3808-3817, 2009.
[http://dx.doi.org/10.1016/j.jmatprotec.2008.08.030]

[17] N. Kaushik, and S. Singhal, "Hybrid combination of Taguchi-GRA-PCA for optimization of wear

behavior in AA6063/SiCp matrix composite", *Prod. Manuf. Res.,* vol. 6, no. 1, pp. 171-189, 2018. [http://dx.doi.org/10.1080/21693277.2018.1479666]

[18] Y.S. Wong, L.C. Lim, I. Rahuman, and W.M. Tee, "Near-mirror-finish phenomenon in EDM using powder-mixed dielectric", *J. Mater. Process. Technol.,* vol. 79, no. 1-3, pp. 30-40, 1998. [http://dx.doi.org/10.1016/S0924-0136(97)00450-0]

[19] S.H. Aghdeab, and A.I. Ahmed, "Effect of pulse on time and pulse off time on material removal rate and electrode wear ratio of stainless steel AISI 316L in EDM", *Eng. Tech. J.,* vol. 35, no. 15 Part A, pp. 2940-2949, 2016.

[20] D. Palanisamy, A. Devaraju, N. Manikandan, K. Balasubramanian, and D. Arulkirubakaran, "Experimental investigation and optimization of process parameters in EDM of aluminium metal matrix composites", *Mater. Today Proc.,* vol. 22, pp. 525-530, 2020. [http://dx.doi.org/10.1016/j.matpr.2019.08.145]

[21] K. Panneerselvam, "Experimental investigation and optimization of abrasive wear characteristics of polypropylene nanocomposites", *Mater. Res. Express,* vol. 7, no. 1, p. 015339, 2020. [http://dx.doi.org/10.1088/2053-1591/ab6b52]

[22] G.S. Balan, and M. Ravichandran, "Study of moisture absorption characteristics of jute fiber reinforced waste plastic filled polymer composite", *Mater. Today Proc.,* vol. 27, pp. 712-717, 2020. [http://dx.doi.org/10.1016/j.matpr.2019.11.260]

[23] S. Kumar, and B.S. Roy, "Novel study of joining of acrylonitrile butadiene styrene and polycarbonate plate by using friction stir welding with double-step shoulder", *J. Manuf. Process.,* vol. 45, pp. 322-330, 2019. [http://dx.doi.org/10.1016/j.jmapro.2019.07.013]

CHAPTER 3

Analyzing the Influence of RSW Process Parameters on the Cross Tension Failure Load of Dissimilar functionally graded AISI 304/316L Stainless Steel Sheets

Vigneshkumar M.[1,*], **Ashoka Varthanan P.**[1], **Jeyakumar R.**[1] and **Raja S.**[1]

[1] *Department of Mechanical Engineering, Sri Krishna College of Engineering and Technology, Coimbatore, Tamilnadu, India - 641008*

Abstract: Joining sheet metals using the spot welding process is the most economical and commonly employed method by the manufacturing industries in producing automotive and aerospace components. This research analyzes the joining of stainless steel of grade EN 1.4301 (AISI 304) welded with EN 1.4435 (AISI 316L) in the cross tension configuration. Four different process parameters, electric current, contact pressure, heating, and squeezing time, are analyzed in the present study. The integrity of the cross lap joints is evaluated by cross tension failure tests and metallurgical analysis. The results are evaluated, and optimal process parameters are obtained by design expert software. A confirmation test is carried out to verify the optimum parameters. It has been found that the electric current is the most influential factor, and a maximum of 14.85 kN cross tension failure load is obtained. The mathematical model developed using the response surface methodology (RSM) has an R2 value of 95.03%, compared with the results of the multilayered ANN model with four neurons in the interface layer. ANN proved to be a slightly better model with an R2 value of 97.6%.

Keywords: ANN, ANOVA, Cross tension strength, Dissimilar sheets, RSW, Stainless steel.

INTRODUCTION

The resistance spot welding process was invented in the eighteenth century. It is still widely used by the automobile and aerospace industries in the fabrication of sheet metals owing to its high manufacturing capability, low cost, and adaptability for automation.

* **Corresponding author Vigneshkumar M.:** Department of Mechanical Engineering, Sri Krishna College of Engineering and Technology, Coimbatore, Tamilnadu, India - 641008; E-mail: vigneshkumar@skcet.ac.in

C. Samson Jerold Samuel, M. Suresh & S. Gnanasekaran (Eds.)

In the resistance spot welding process, the workpieces to be welded are held between two copper electrodes under pressure for the set squeeze time. After that, a large amount of electric current is passed between the electrodes while maintaining the pressure for a short time called heating time. As per Ohms law, the contact resistance between the two sheets causes heat generation at the interface. It causes localized melting and solidification at the faying surfaces. The pressure is maintained after welding for a short period. After that, the electrodes are withdrawn, and the next set of welding cycles is continued.

Automobile industries are looking for the inclusion of lightweight, high strength, and corrosion resistance structures in their assembly to meet the cost and energy-efficient demands [1]. Joining dissimilar metals is one way of meeting the high demands as it delivers tailor engineered properties to the assembly. The higher electrical resistivity and mechanical properties of austenitic stainless steel make it more suitable for the manufacturing vehicle assembly by the resistance spot welding process [2]. Austenitic stainless steel of grade EN 1.4435 provides superior corrosion resistance and good mechanical properties owing to the alloying content such as molybdenum, chromium, and nickel and lower carbon content of 0.03%. EN 1.4435 stainless steel sheets are being used in coastal areas for marine applications. The chemical industries are also using it for making pipes and in biomedical applications [3]. Thus, the joining of EN 1.4301 sheets with EN 1.4435 provides the assembly of higher corrosion resistance with better mechanical properties at a lower cost, as EN 1.4301 is inexpensive.

The crashworthiness of an automobile is an important factor that describes its ability to safeguard its passengers at the time of an accident. It is governed by the integrity and strength of spot-welded joints.Spot welds in automobiles are subjected to shear and tensile loading in their operation. Shear loads at the spot welds are induced by sheet movement concerning another sheet or sheet rotation. The separating forces between the two sheets in the normal direction cause tensile loading on spot welds. The tensile shear failure test is primarily used to analyze the quality of resistance spot welded joints on shear loading, while the cross tension test is used for tensile loading. Many researchers analyzed the mechanical strength of similar and dissimilar resistance spot-welded joints based on the tensile shear failure test. Zhao *et al.* [4] investigated the resistance spot weldability of TC2 titanium alloy. They found that the peak tensile shear failure load of 2.45 kN is attained at 2.4 kA of electric current and 9.39 milliseconds of heating time. The tensile shear failure load of dissimilar advanced high strength steels (B250P1 & B1500HS) is 4.5 kN at 7.1 kA of electric current and 40 milliseconds of heating time [1]. Zhang *et al.* [5] analyzed the dual-phase steels spot weldability with DP780 and DP600 sheets. The peak failure load is found to be around 20 kN under tensile shear. They found that crack propagation is initiated in the heat-affected zone of the DP780 side due to the stress concentration and softening of the heat-affected zone. Tamizi *et al.* [6] studied the resistance spot welding of martensitic advanced high strength stainless steel. They observed less hardness in the HAZ than the base metal because of the ferrite's formation and tempering of martensite. They also suggested that minimum welding heat input is required to obtain adequate fusion zone size for the pull-out mode of failure in the tensile shear failure test. Moshayedi *et al.* [7] analyzed the influence of electric current and heating time on stainless steel spot weldability. They constructed a computational 2D model for studying the growth of weld nuggets. They concluded that heating time has little effect on nugget size compared to welding current. Sun *et al.* [8] attempted to join aluminium and magnesium sheets with the help of a nickel interlayer to avoid the development of magnesium – aluminium fusible compounds. They observed a considerable increase in the weld strength due to the presence of an interlayer.

Few reports are found on the analysis of the integrity of the spot welds based on cross tension tests. Chao *et al.* [9] analyzed the cross tension failure load of alloy steels. They found that the cross tension failure load lies between 2.3 kN and 4 kN with the variation of electric current and heating time. In contrast, the failure load value lies between 5.63 kN and 4.63 kN under tensile shear for different parameter levels. Tozaki *et al.* [10] studied the cross tension strength of friction stir spot welded sheets. They found that the peak cross tension failure load of 4 kN is attained for the aluminium alloy, and it decreases with the increase of tool rotational speed. Welding of dissimilar metals is difficult compared to similar metals due to the difference in material properties of the base metals [11]. The literature shows that several extensive studies are carried out on the resistance spot welding of austenitic, ferritic, and martensitic steels based on the tensile shear failure load [12 - 18]. However, the cross tension failure load of resistance spot welded dissimilar metals are little explored. The cross tension failure load is found to be lesser than the tensile shear failure load in most case studies, and it must be considered in the analysis of the crashworthiness of an automobile. Hence, the objective of the current study is to spot weld EN 1.4301 grade stainless steel with EN 1.4435 stainless steel in the cross tension configuration and analyze the effect of process parameters on the quality of the spot-welded joints.

EXPERIMENTAL STUDIES

Specimen Preparation and Welding

A 35KVA resistance spot welding system is used in this research to weld stainless steel EN 1.4301 and EN 1.4435. The machine is capable of setting the contact pressure pneumatically. The truncated copper electrode used in this machine belongs to Group A of Resistance welder manufacturing alliance (RWMA) with a diameter of 7 mm, and it is water-cooled. Specimens are configured for cross tension failure testing and microstructural analysis as per AWS D8.9-12 standard. Using a mechanized shearing machine, specimens are cut to the appropriate size of 150 mm × 50 mm for cross tension testing and 30 mm × 30 mm for microstructural analysis. Preliminary experiments are conducted to identify practical process parameter limits for the RSW of selected materials. The weld is located in the middle of the plates at W/2 and L/2, where the width and length of the plates are W and L. The spot-welded cross tension test specimens are shown in Fig. (**1**). The selected process parameter levels are shown in Table **1**. The comprehensive analysis is carried out through experiments based on Central Composite Design (CCD). The process parameters for this analysis include contact pressure, electrical current, heating time and squeeze time.

Fig. (1). AISI 304/316L Spot welded Cross tension test specimens.

Testing of Specimens

The cross tension strength of the spot-welded joints is measured using 50 kN universal testing equipment. A fixture is prepared as per AWS D8.9M -12 standards to minimize the bending of specimens, as shown in Fig. (**2**). The fixture consisted of two separate units connected using a nut and bolt. Specimens are gripped between these two units. The fixtures have a cylindrical rod welded to the other end of the gripping surfaces. The cylindrical rods are attached to the universal testing machine for the application of load. Four holes of 20 mm diameter are drilled on the workpiece for gripping it using the fixture. The specimen's width is 50 mm, and the hole spacing HS is kept as 100 mm. During the application of load, specimens are subjected to normal forces at the interface. The crosshead velocity is maintained at 10 mm/min as per the ASTM D8.9M - 2012 standard. The results of the cross tension test are shown in Table **2**. For the microstructural analysis, 30 mm × 30 mm is prepared and welded at the centre. The welded specimens are sectioned and polished using an Al_2O_3 abrasive wheel with a grain size of 0.05 µm. An aqueous solution (25 ml HNO_3 + 50 ml HCl+ 25 ml H_2O) is used for the etching process, and microstructures are analyzed using Sigma HV Field emission Scanning electron microscope (FSEM).

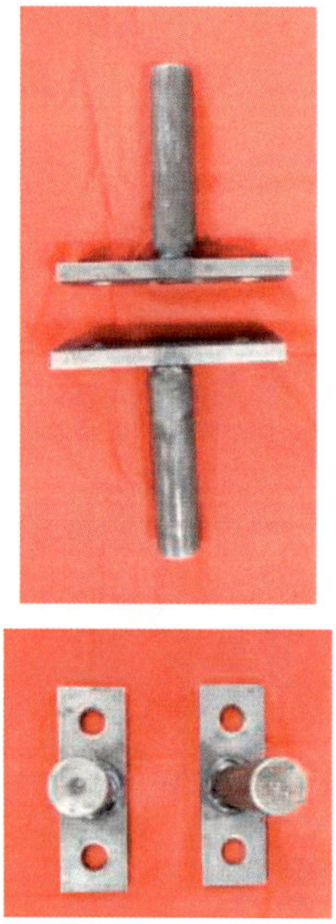

Fig. (2). Fixtures of cross tension test.

MATHEMATICAL MODELLING

RSM Modelling

An empirical model is developed using response surface methodology through design expert software. Four spot welding parameters such as contact pressure, electric current, heating time, and squeeze time are considered in the present study. These four factors are varied at five different levels based on axial, central, and factorial points of central composite design. Experiments are conducted at different levels at each process parameter, based on CCD Design. Experiments are repeated six times at the central points for precision modelling. The results of properly designed experiments are used to generate a regression equation. The general second-order polynomial equation for four factors is shown below in equation 1.

$$y = \beta_0 + \sum_{i=1}^{4}(\beta_i)x_i + \sum_{i=1}^{4}((\beta_{ii})x_i^2 + \sum_{i=1}^{4}\sum_{j=i+1}^{4}((\beta_{ij})x_i x_j) \quad (1)$$

Where Y is the predicted cross tension failure load, $\beta_{0,}$ $\beta_{i,}$ β_{ii}, and β_{ij} is constant, linear quadratic, and cross-product coefficient, respectively. The term $x_i x_j$ denoted the interaction between the process parameters.

Artificial Neural Network Modelling

An artificial neural network is a computing system that consists of several layered structures that connect output and input. It resembles our human brain, which is interconnected by several neurons. Neural networks can retrieve information from the data for learning, and they can be trained to predict the outcomes of a variable. Neurons are used for processing information in ANN. The neurons are connected with interlinks or connection links. These connection links possess a weight that will multiply with the incoming signal for calculating each node's output. In a multilayered structure, the input layer is connected to the output layer through several hidden layers for processing the signals. The function of neurons in the hidden layer is based on input neurons' activity and the weightage assigned between the interlinks of input and hidden layer. Training operation in ANN enables assigning or modifying the weights in the connection to achieve the desired output. Non-linear functions are preferred in a multilayered net. Examples of non-linear functions are sigmoidal functions. The weights of each input are given in the matrix form, and then the net input is calculated using matrix multiplication as below in equation 2.

$$y_{input} = \sum X_i W_{ij} \quad (2)$$

For i = 1 to n, where n is the number of inputs.

The bias improves the accuracy of the ANN model by altering its output. The error function is calculated and incorporated into the input function as bias, as shown in equation 3.

$$Net\ Input = \sum X_i W_{ij} + bias \quad (3)$$

RESULTS AND DISCUSSIONS

The experimental results of the cross tension test based on response surface methodology central composite design of experiments are analyzed using the Design-Expert software 10.1 version. A quadratic equation is obtained with cross tension failure load as a function of four process parameters: contact pressure, welding current, Heating time, and squeezing time. The quadratic equations in terms of coded and actual values are shown below in equation 4. In terms of coded factors as listed in Table **1**.

Cross tension failure load (Y) = +12.88 -0.14A +0.54B+0.39C+0.13D-0.44AB-0.37AC+0.022AD-0.45BC -0.14 BD -0.12CD+0.020A^2 -0.092 B^2 -0.34 C^2 -0.1D^2 (4) **(4)**

Where Y = Cross tension shear failure load (kN), A = Contact pressure, B = Electric current, C = heating time & D= Squeezing time.

ANOVA for Determining the Influence of Process Parameters

An Analysis of Variance (ANOVA) accesses the impact of each process parameter and the validity of the established mathematical model. The ANOVA results for cross tension strength are shown in Table 3. The F statistics output of the model is found to be 20.47, and the probability value $>F<0.0001$ indicates that the developed model is significant in predicting cross-tension strength. The lack of fit value of 0.0566 suggests that inadequate fit is negligible in the established mathematical model. Among the selected four process parameters, two parameters, namely electric current and heating time, are significant with P value>F and less than 0.0001. The other two process parameters, contact pressure and squeeze time, are found to be insignificant. The interaction between contact pressure and electric current, electric current, heating time and quadratic terms of heating time is significant in the influence of cross tension failure load. The other interaction factors, such as contact pressure and heating time, contact pressure and squeezing time, electric current and squeezing time, heating time, and squeezing time, are found to be insignificant. Among the selected four process parameters, the electric current is the most dominant process parameter. The lack of 0.0566 fit value showed that it is not important in relation to the pure error. The R^2 value estimates the accuracy of the developed mathematical model. At a 95% confidence level, the developed model has an R^2 value of 95.03. It shows that a 95.03% variation in the output response (Cross tension failure load) is due to the studied process parameters, and the noise in the model amounts to only 4.97%.

Effect of Process Parameters

The effect of four process parameters such as contact pressure, welding current, heating time, and squeeze time on the cross tension failure load are analyzed with three-dimensional surfaces developed using design expert software based on the RSM. The effect of contact pressure and squeeze time on cross tension strength is shown in Fig. (**3a-c**). At constant electric current and heating time of 8 kA and 8 cycles, respectively, the 3D diagram is plotted. At 1.5 bar of contact pressure and 10 cycles of heating time, the cross tension failure load is 12.8 kN. As the pressure increases to 2.5 bar, the cross tension failure load decreases to 12.1 kN owing to the reduction in contact resistance between the specimen surfaces. The 3D diagram could be inferred that squeeze time had a flat variation over cross tension strength. At 2.5 bar of contact pressure, the cross tension failure load is 12.1 kN and 13 kN at 10 cycles and 13 cycles, respectively. Compared with other process parameters, both the squeeze time and contact pressure had little influence on cross tension strength.

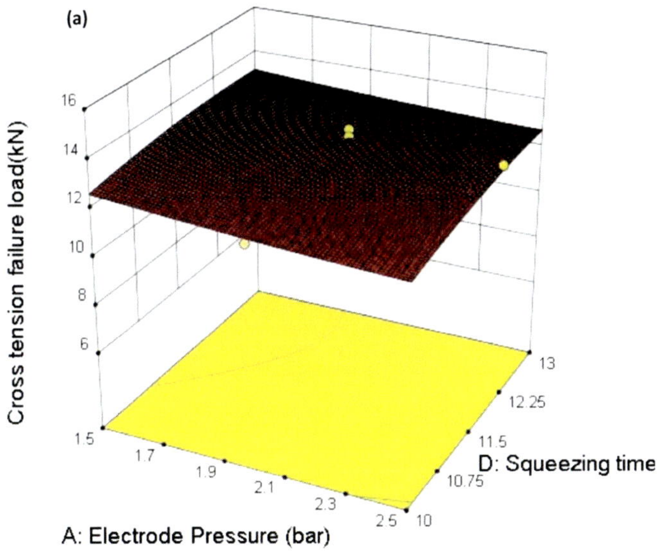

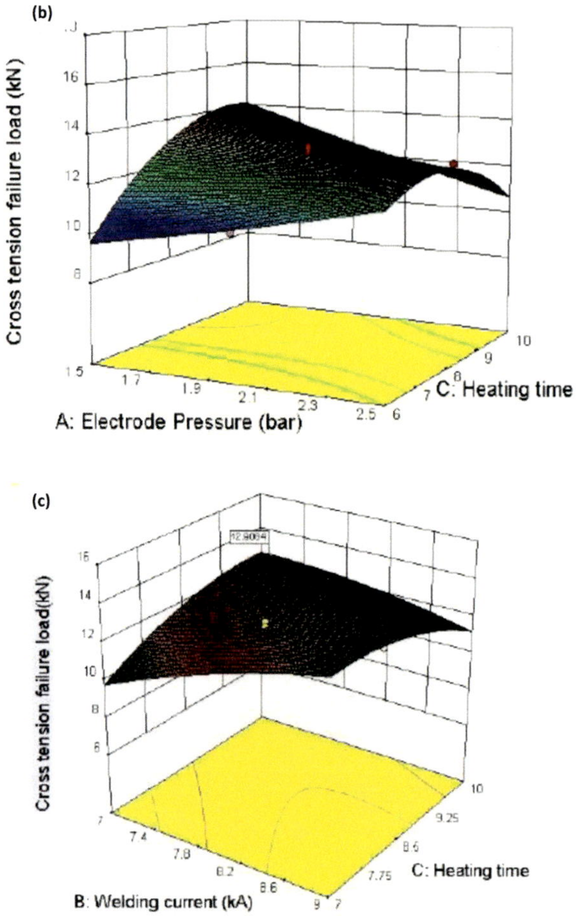

Fig. (3a-c). Effect of process parameters on cross tension failure load.

The effect of contact pressure and heating time on cross tension failure load at constant electric current and squeeze time of 8.25 kA and 12 cycles are illustrated in Fig. **3(b)**. It is found that the influence of contact pressure is less when compared to heating time. At 1.6 bar of contact pressure and 6 cycles of heating time, the cross tension failure load is 11.1 kN. The weld strength is found to be low due to insufficient heat input. As the heating time increases to 8 cycles with other process parameters at the same levels, the cross tension strength increases to 14.01 kN, and the further increase of heating time increases the cross tension strength to 14.4 kN at 9 cycles of heating time. Beyond 9 cycles of heating time, the cross tension strength starts decreasing due to the expulsion of molten metal because of overheating. At higher contact pressures, the cross tension failure load is found to be low. At 2.5 bar of contact pressure, the peak cross tension failure load of 12.75 kN is observed at 8 cycles of heating time. The decrease of cross tension strength at higher contact pressure is attributed to a decrease in contact resistance.

The effect of electric current and heating time on the cross tension strength is depicted in the 3D diagram, as shown in Fig. **3(c)**. The 3D diagram is developed at constant contact pressure and squeezing time of 2 bar and 12 cycles, respectively. At 9 kA of electric current and 7 cycles of heating time, the cross tension strength is found to be 14 kN, whereas, at 10 kA of electric current and 10 cycles of heating time, the cross tension strength is found to be 11.5 kN. The decrease of cross tension strength at higher values of both the electric current and heating time is due to overheating, causing the ejection of molten metal from the faying surfaces. The cross tension strength is maximum when the electric current and heating time are moderate, indicating optimum heat input.

When the load is applied to the cross tension specimen, the sheets are initially exposed to bending stresses, creating considerable bending deformation. The mode of failure of the specimen is observed to be bending and thickness-shear of the weld nugget. The initial 40 mm crosshead travel for a 5 kN load is attributed mainly to the bending deformation. The subsequent increase of load causes the weld nugget to be pulled out of one of the specimens as depicted in the fracture specimen, as shown in Fig. **(4)**. Beyond 40 mm crosshead travel, bending and shearing of the weld nugget are observed with the maximum failure load of 14.9 kN.

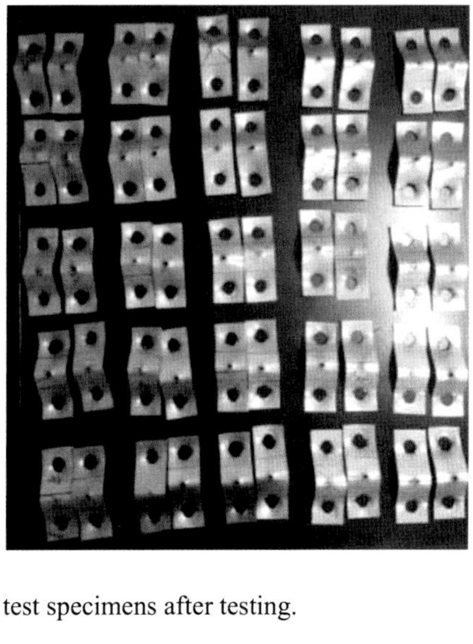

Fig. (4). Fractured cross tension test specimens after testing.

Confirmatory Experiments at Optimum Process Levels

Using the numerical optimization method of the Design–Expert v12 software, the optimized process parameters and levels for higher cross tension strength are identified. The objective is to maximize the cross tension failure load, and the solution with desirability 1 is found to be 15.2 kN at 1.56 bar of contact pressure, 8.7 kA of electric current, and heating time of 9 cycles. The mathematical results are validated at the closest possible machine settings by the confirmatory test (Contact pressure – 1.55 bar, Electric current – 8.7 kA, Heating time – 9 cycles, Squeeze time – 12 cycles). Two specimens are tested at the optimized process parameter settings, and the results are shown in Table 4. The load *versus* crosshead travel plot for the confirmation tests is shown in Figs. (**5a** and **5b**). The cross tension test failure load for the confirmation test 1 and 2 is 14.9 kN and 14.8 kN. The mode of failure is found to be bending and shearing. The cross tension failure load of 14.9 kN is 5.67% higher than the maximum cross tension failure load of 14.1 kN obtained during the experimental analysis.

(a)

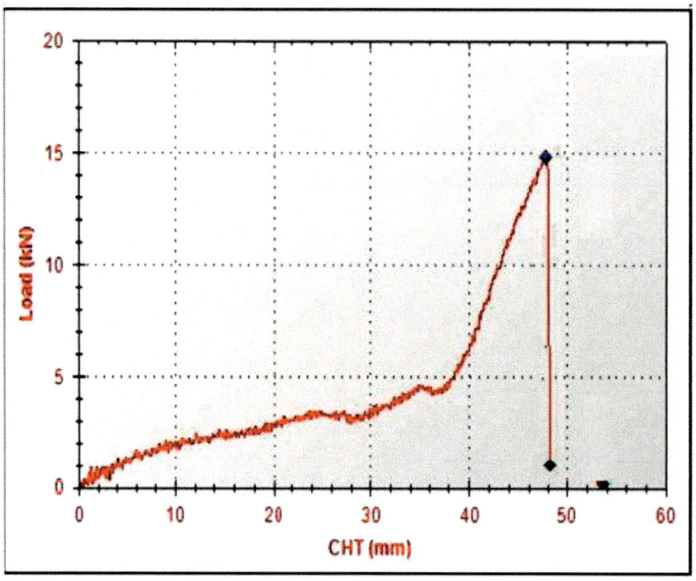

(b)

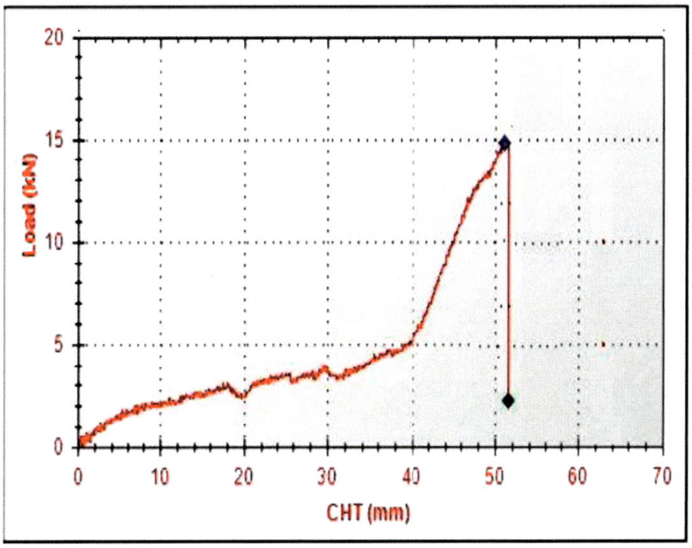

Fig. (5 a and b). Load *vs* cross head travel plot for the confirmation tests.

Artificial Neural Network Analysis of Cross Tension Load

An artificial neural network model is developed using MATLAB 9.4 Software with four input neurons to analyze the effect of process parameters on the cross tension failure load. The output layer is connected with four input neurons through a hidden layer. Each input neuron represents each process parameter, like contact pressure, welding current, heating time, and squeeze time. A feed-forward backpropagation algorithm with a sigmoid transfer function is used in the current analysis. The selected transfer function is differentiable and could be used to determine the neural network's output from its input of the four neurons. Based on the input values, weights, and biases, the output of the neuron is calculated. The weights and biases are updated during the training operation. For training the network, traingdx function with gradient descent momentum is considered. The mean square error is calculated for different topologies with different hidden layer neuron counts, and the neuron count corresponding to the minimum mean square error is selected [19]. The mean square error is determined based on the experimental and ANN predicted results, based on the following equation 5.

$$MSE = \frac{1}{N}\sum_{i=1}^{N} [(S_{ANN})_i - (S_{exp})_i]^2 \quad (5)$$

Whereas $(S_{exp})_i$ and $(S_{ANN})_i$ are the experimental and ANN predicted results' respectively. In the current study, mean square error (MSE) is found to vary between 0.0023 to 0.040 for different topologies when the hidden layer neuron counts are varied between 1 and 10 as shown in Fig. (**6**).

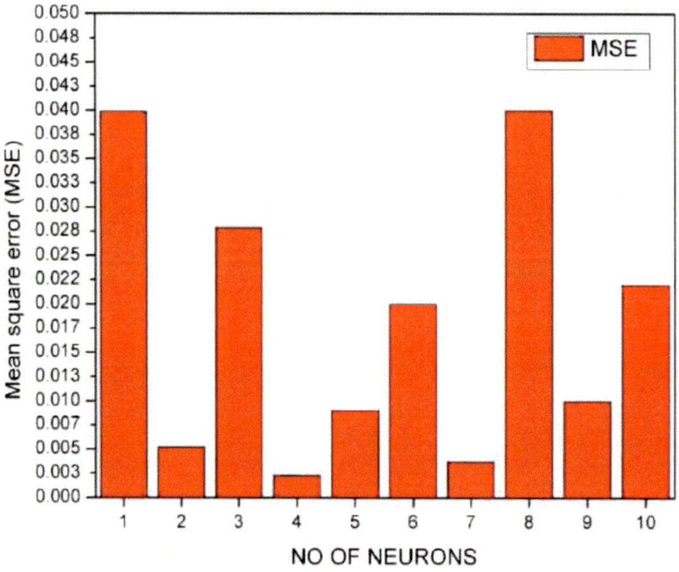

Fig. (6). Effect of number of neurons in the hidden layer on the mean square error (MSE).

MSE is calculated for different topologies with a varying number of hidden layers to select the best topology. Mean square error is found to be in the range of 0.0023 to 0.040 for different topologies. The minimum mean square error value of 0.0023 is observed for the hidden layer neuron count of four, as shown in Fig. (**7**), and is selected for the analysis. Thus, a multilayered architecture with four hidden layer neurons is considered to predict cross tension strength, as shown in Fig. (**8**). For the four neurons in the hidden layer, the weights and biases are optimized for minimum mean square error (MSE) using traingdx function.

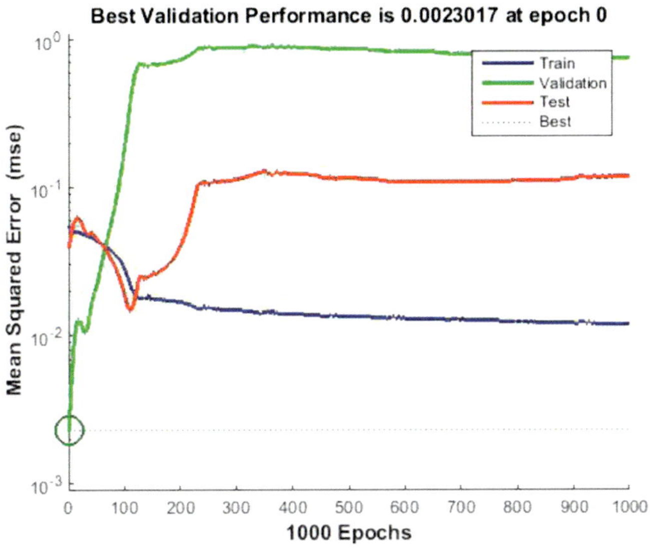

Fig. (7). Minimum mean square error (MSE) for 4 neurons in the hidden layer.

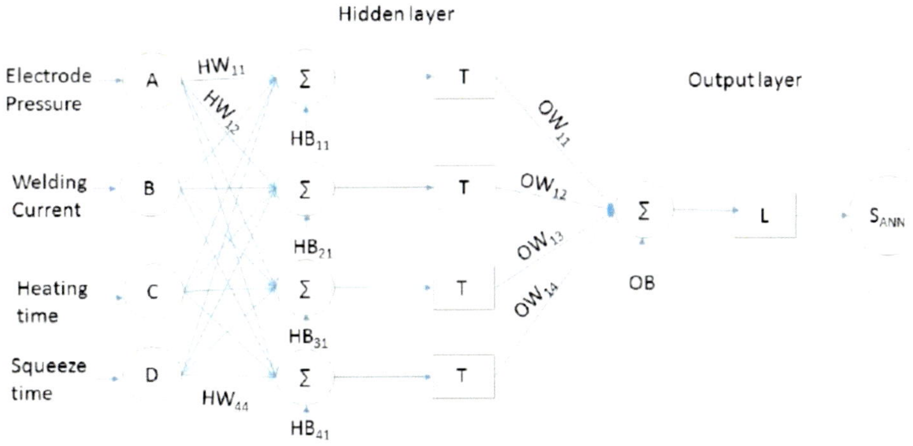

Fig. (8). ANN Architecture for the prediction of cross tension strength

The cross tension strength values are predicted with optimized weights and biases of each neuron, and its values for each experiment are shown in Table **5**. RSM and ANN predicted values of cross tension strength are shown in Table **2**. The ANN model's R2 value in predicting the cross tension strength is found to be 97.65%, as shown in Fig. (**9**). The average absolute error value and mean absolute deviation for RSM predicted cross tension model is found to be 0.284 kN and 0.1544 kN, respectively. The maximum error value of the RSM and ANN model is found to be 0.79 kN and 0.69 kN, respectively. The average absolute error value and mean absolute deviation for ANN predicted cross tension model is found to be 0.126 kN and 0.127 kN, respectively. It shows that the results of the ANN model agreed with both the experimental and RSM results.

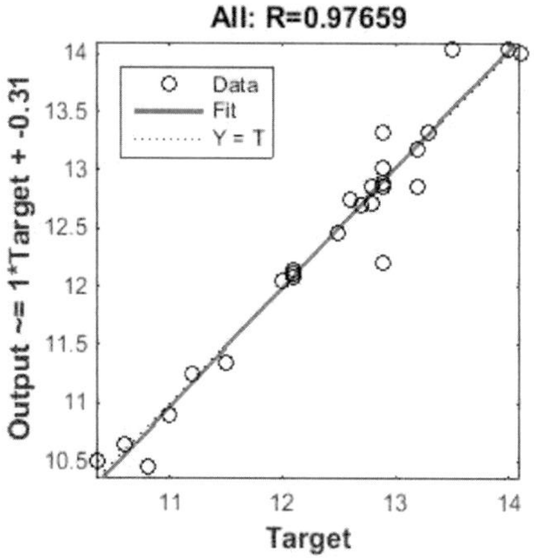

Fig. (9). R^2 value of the ANN model for predicting the cross tension strength.

SEM Analysis for Analyzing the Weld Nugget

The microstructure of the specimen at optimum process parameter levels of 8.7 kA of electric current and 9 cycles of weld time is analyzed using Scanning Electron microscope analysis(SEM), as shown in Fig. (**10a-e**). The microstructure of the entire weld nugget is shown in Fig. (**10a**). The weld's microstructure is free of voids and intergranular crack in the heat-affected zone because of excess heat input, indicating a sound weld joint. The microstructure of the base metals away from the weld nugget is shown in Fig. (**10b**). The centerline indicates the gap between two sheets, and it decreases in the direction of the weld nugget. Both EN 1.4301 and EN 1.4435 are found to have austenitic microstructure. The alloying elements in the EN 1.4435 are higher than EN 1.4301. The microstructure of base metals and weld nugget is shown in Fig. (**10c**). Both the base metals melted and solidified during the formation of weld nuggets. The gap between the two sheets is not observed near the weld nugget. The microstructures of the specimens at the interior of the weld nugget and near the boundary are shown in Fig. (**10d-e**), respectively. In the weld nugget interior, the microstructure consists of austenite, whereas near the boundary, it comprises ferrite and austenite structure. It is attributed to the solidification of base metals in the kinetic austenite ferrite mode.

Because of the existence of water-cooled electrodes, the cooling rate is higher at the boundary. Thus, the microstructure indicated successful spot welding of EN 1.4301/EN 1.4435 stainless steel sheets free of intergranular crack and voids. The microstructure observed that EN 1.4301/EN 1.4435 fusion zone is symmetrical, and the final solidification line coincided with the sheet interface. It is attributed to the fact that the electrical resistivity of both metals is nearly the same. For the symmetrical fusion zone, the dilution value for both the base metals is 50%. The dilution ratio is the proportion of melted base metal volume to the total weld nugget volume. Nugget offset is not observed as the dilution ratio is 50%, which is highly desired for the better mechanical performance of spot-welded joints.

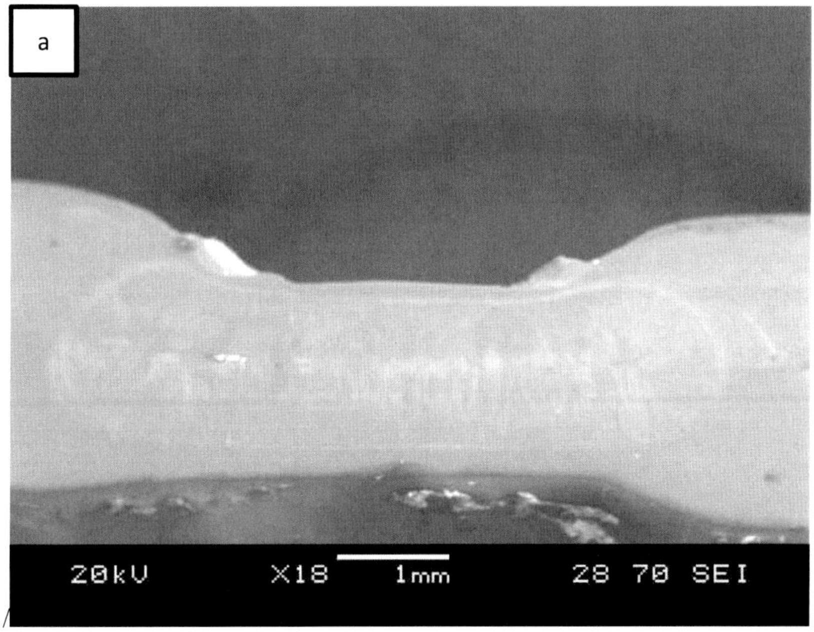

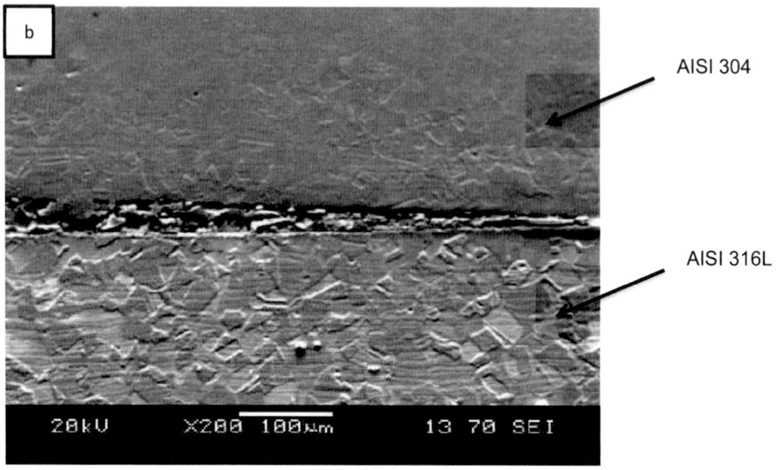
b
AISI 304
AISI 316L
20kV X200 100µm 13 70 SEI

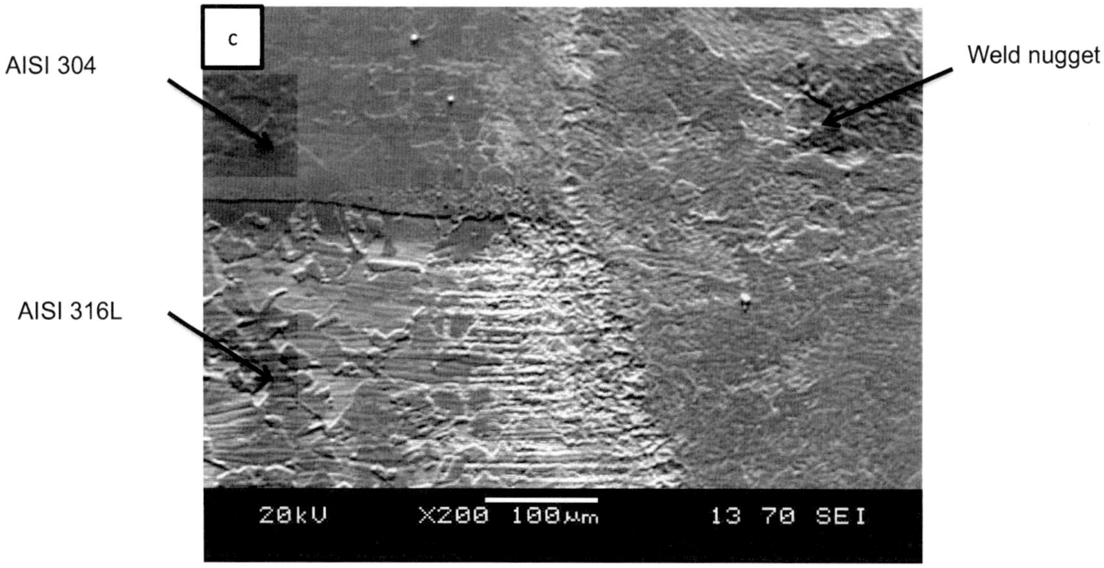
c
AISI 304
Weld nugget
AISI 316L
20kV X200 100µm 13 70 SEI

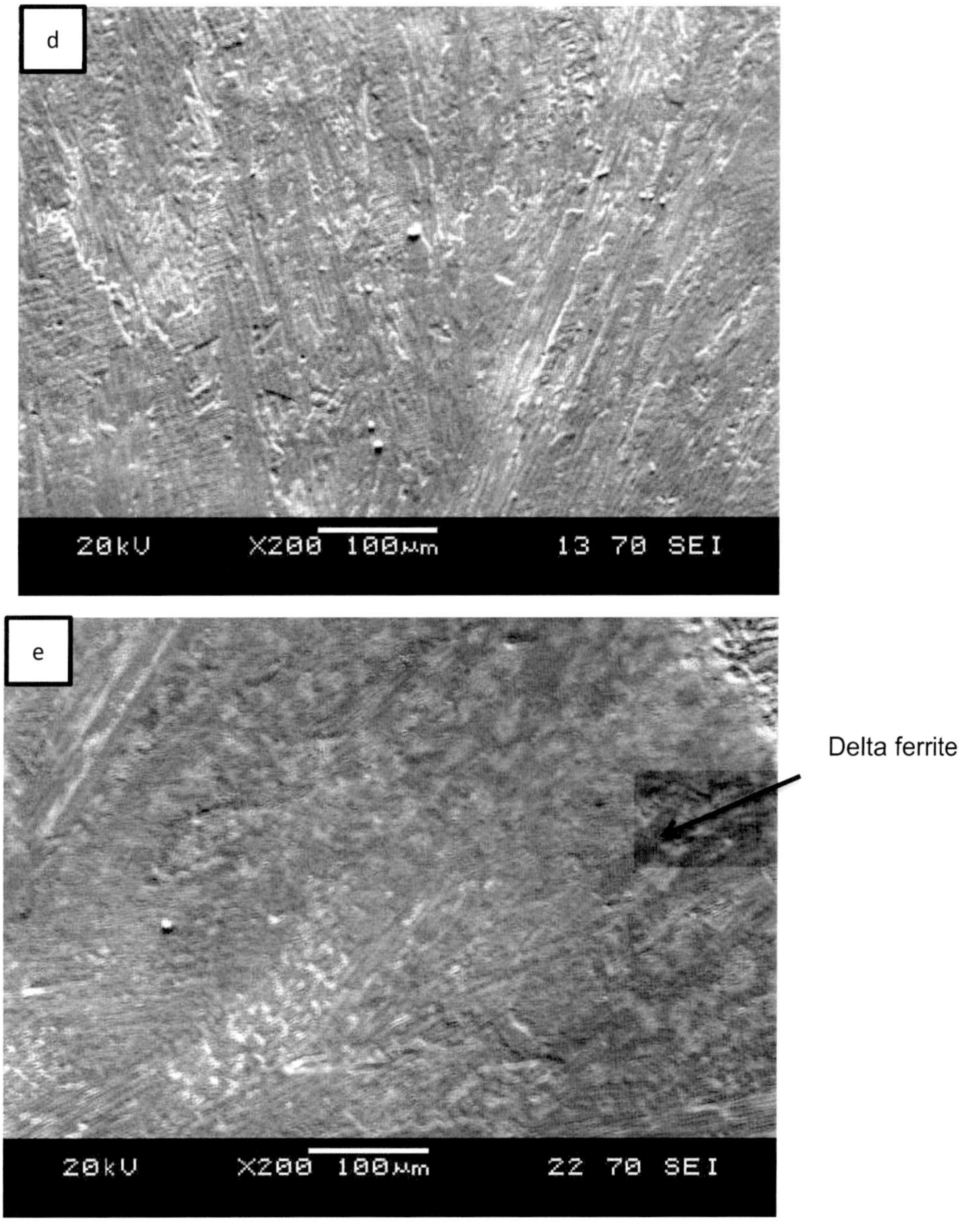

Fig. 10(a-e). SEM images of the AISI 304/316L spot welded specimen at 8.7 kA welding current and 9 cycles welding time.

CONCLUSION

i. The experimental investigation found that electric current and heating time are the dominant factors influencing the cross tension failure load.

ii. The maximum cross tension failure load of 14.9 kN is obtained at an optimal process parameter setting, which is 5.67% higher than the cross tension failure load value of 14.1 kN obtained during the experimental runs.
iii. The cross tension specimens are subjected to a large amount of bending deformation with crosshead travel of up to 50 mm, followed by subsequent nugget shearing. Thus, the mode of failure is found to be bending and shearing the weld nugget.
iv. A multilayered ANN architecture is utilized to predict the cross-tension strength. The average absolute error value of the RSM and ANN-based cross tension model is found to be 0.284 kN and 0.126 kN, respectively. The maximum error of the RSM and ANN model is found to be 0.79 kN and 0.69 kN.
v. Both the RSM and ANN models provided quality predictions in determining tensile shear failure load and cross tension load. ANN model is found to be slightly better than the RSM model with a higher R^2 value.
vi. Based on the microstructural analysis, nugget offset and voids are not observed, and it is desired for better mechanical performances of welded joints. Fractography reveals high stress with plastic deformation caused by the nucleation of voids and localized strain during failure.

CONSENT FOR PUBLICATION

Not applicable.

CONFLICT OF INTEREST

The author declares no conflict of interest, financial or otherwise.

ACKNOWLEDGEMENTS

Declared none.

REFERENCES

[1] Y. Luo, W. Rui, X. Xie, and Y. Zhu, "Study on the nugget growth in single-phase AC resistance spot welding based on the calculation of dynamic resistance", *J. Mater. Process. Technol.,* vol. 229, pp. 492-500, 2016.
[http://dx.doi.org/10.1016/j.jmatprotec.2015.10.006]

[2] W.H. Kearns, "Welding handbook, Vol. 4, Metals and their weldability", In: *American Welding Society*, 1982, no. 1982, p. 582.

[3] T. Jagadeesha, and T.J. Sarvoththama Jothi, "Studies on the influence of process parameters on the EN 1.4435 resistance spot-welded specimens", *Int. J. Adv. Manuf. Technol.,* vol. 93, no. 1, pp. 73-88, 2017.

[4] D. Zhao, Y. Wang, S. Sheng, and Z. Lin, "Multi-objective optimal design of small scale resistance spot welding process with principal component analysis and response surface methodology", *J. Intell. Manuf.,* vol. 25, no. 6, pp. 1335-1348, 2014.

[http://dx.doi.org/10.1007/s10845-013-0733-2]

[5] H. Zhang, A. Wei, X. Qiu, and J. Chen, "Microstructure and mechanical properties of resistance spot welded dissimilar thickness DP780/DP600 dual-phase steel joints", *Mater. Des.*, vol. 54, pp. 443-449, 2014.
[http://dx.doi.org/10.1016/j.matdes.2013.08.027]

[6] M. Tamizi, M. Pouranvari, and M. Movahedi, "Welding metallurgy of martensitic advanced high strength steels during resistance spot welding", *Sci. Technol. Weld. Join.*, vol. 22, no. 4, pp. 327-335, 2017.
[http://dx.doi.org/10.1080/13621718.2016.1240979]

[7] H. Moshayedi, and I. Sattari-far, "Numerical and experimental study of nugget size growth in resistance spot welding of austenitic stainless steels", *J. Mater. Process. Technol.*, vol. 212, no. 2, pp. 347-354, 2012.
[http://dx.doi.org/10.1016/j.jmatprotec.2011.09.004]

[8] H. Sun, J. Yang, and L. Wang, "Resistance spot welding quality identification with particle swarm optimization and a kernel extreme learning machine model", *Int. J. Adv. Manuf. Technol.*, vol. 91, no. 5, pp. 1879-1887, 2017.
[http://dx.doi.org/10.1007/s00170-016-9944-y]

[9] Y.J. Chao, and J. Eng Mater Technol, "Ultimate Strength and Failure Mechanism of Resistance Spot Weld Subjected to Tensile, Shear, or Combined Tensile/Shear Loads", *J. Eng. Mater. Technol.*, vol. 125, no. 2, pp. 125-132, 2003.
[http://dx.doi.org/10.1115/1.1555648]

[10] Y. Tozaki, Y. Uematsu, and K. Tokaji, "Effect of tool geometry on microstructure and static strength in friction stir spot welded aluminium alloys", *Int. J. Mach. Tools Manuf.*, vol. 47, no. 15, pp. 2230-2236, 2007.
[http://dx.doi.org/10.1016/j.ijmachtools.2007.07.005]

[11] M.H. Bina, M. Jamalina, M. Shamanian, and H. Sabet, "Effect of Heating time in the Resistance Spot Welded Dissimilar Stainless Steels", *Trans. Indian Inst. Met.*, vol. 68, no. 2, pp. 247-255, 2015.
[http://dx.doi.org/10.1007/s12666-014-0452-1]

[12] P. Spena, F. De Maddis, F. Lombardi, and M. Rossini, "Dissimilar Resistance Spot Welding of Q&P and TWIP Steel Sheets", *Mater. Manuf. Process.*, vol. 31, no. 3, pp. 291-299, 2016.
[http://dx.doi.org/10.1080/10426914.2015.1048476]

[13] S.T. Wei, R.D. Liu, D. Lv, L. Lin, X.F. Lu, and S.W. Ding, "Effect of joint configuration on resistance spot weldability of galvanised DP780 steel sheets", *Sci. Technol. Weld. Join.*, vol. 21, no. 3, pp. 178-185, 2016.
[http://dx.doi.org/10.1179/1362171815Y.0000000081]

[14] M. Alizadeh-Sh, S.P.H. Marashi, and M. Pouranvari, "Resistance spot welding of AISI 430 ferritic stainless steel: Phase transformations and mechanical properties", *Mater. Des.*, vol. 56, no. 1, pp. 258-263, 2014.
[http://dx.doi.org/10.1016/j.matdes.2013.11.022]

[15] D.S. Safanama, S.P.H. Marashi, and M. Pouranvari, "Similar and dissimilar resistance spot welding of martensitic advanced high strength steel and low carbon steel: metallurgical characteristics and failure mode transition", *Sci. Technol. Weld. Join.*, vol. 17, no. 4, pp. 288-294, 2012.
[http://dx.doi.org/10.1179/1362171812Y.0000000006]

[16] M. Vigneshkumar, P. Ashoka Varthanan, and Y. Maria Ambrose Raj, "Finite Element-Based Parametric Studies of Nugget Diameter and Temperature Distribution in the Resistance Spot Welding of EN 1.4301 and EN 1.4435 Sheets", *Trans. Indian Inst. Met.*, vol. 72, no. 2, pp. 429-438, 2019.
[http://dx.doi.org/10.1007/s12666-018-1494-6]

[17] M. Vural, and A. Akkus, "On the resistance spot weldability of galvanized interstitial free steel sheets with austenitic stainless steel sheets", *J. Mater. Process. Technol.*, vol. 153, pp. 1-6, 2004.

[http://dx.doi.org/10.1016/j.jmatprotec.2004.04.063]

[18] K. Kishore, P. Kumar, and G. Mukhopadhyay, "Resistance spot weldability of galvannealed and bare DP600 steel", *J. Mater. Process. Technol.,* vol. 271, pp. 237-248, 2019. [http://dx.doi.org/10.1016/j.jmatprotec.2019.04.005]

[19] A.K. Mahamad, S. Saon, and T. Hiyama, "Predicting remaining useful life of rotating machinery based artificial neural network", *Comput. Math. Appl.,* vol. 60, no. 4, pp. 1078-1087, 2010. [http://dx.doi.org/10.1016/j.camwa.2010.03.065]

CHAPTER 4

A Review on Manufacturing of Aluminium Alloy Based Metal Matrix Composites through Stir Casting Route

S.N. Vijayan[1,*], Samson Jerold Samuel Chelladurai[2], Saiyath Ibrahim A.[1], R. Ramesh Babu[1], P. Prakash[1] and **S. Gnanasekaran[3]**

[1] *Department of Mechanical Engineering, Karpagam Institute of Technology, Coimbatore, TamilNadu, India*

[2] *Department of Mechanical Engineering, Sri Krishna College of Engineering and Technology, Coimbatore, TamilNadu, India*

[3] *Department of Mechanical Engineering, Sri Sakthi Institute of Engineering and Technology, Coimbatore, TamilNadu, India*

Abstract: A lot of research has been carried out over the past decades to develop an appropriate material that meets the rising demands of the industries. Aluminium is one of the most commonly used lightweight materials because of its corrosion resistance, soft ductility, and excellent electrical conductivity. However, aluminum is not efficient in high thermal loads due to its soft nature. Metal Matrix Composites (MMCs) are being built to address this drawback, which is a new class of engineering materials in which aluminum alloys are being used as a matrix. Hard ceramic particles (SiC, B_4C, Al_2O_3 & ZrB_2) are generally used as reinforcement phases. Such MMCs based on aluminum have enhanced mechanical properties such as high strength-to-weight ratio, better wear resistance, good fatigue resistance, and outstanding corrosion resistance, making them ideal for use in various fields such as marine, automotive, aerospace, and defense. Among different casting technologies, stir casting technology may be used because of its easy fabrication method and low manufacturing cost.

Keywords: Alloy, Aluminium, Hardness, Light weight material, Metal Matrix Composite, Reinforcement Particle, Stir Casting, Tensile, Tribology and Mechanical properties, Wear resistance.

INTRODUCTION

A variety of work is carried out on regular demand for advanced materials in the engineering industries that have lightweight and high strength. The lightweight

* **Corresponding author Vijayan.S.N:** Assistant Professor, Department of Mechanical Engineering, Karpagam Institute of Technology, Coimbatore, India; E-mail: s.n.vijayan@gmail.com.

C. Samson Jerold Samuel, M. Suresh & S. Gnanasekaran (Eds.)

aluminum material is considered fit for all engineering applications due to its multifunctionality, simplicity and high performance [1].

Aluminium alloy ingots are generally classified based on the British standard 1490:1988 as group A, group B, and group C. Group A contains Alloys of LM2, LM4, LM6, LM20, LM24, LM25 and LM27. Because of their good characteristics, these materials are commonly used for all forms of die castings and can increase strength and hardness during the heat treatment process. Group B considers special purpose alloys that are LM0, LM5, LM9, LM13, LM16, LM21, and LM22. These special purpose alloys are frequently used for sand casting and chill casting processes requiring high corrosive resistance characteristics. These materials are used for manufacturing electrical and chemical components used in food processing industries and marine applications. These alloys tolerate high thermal stresses and have mechanical properties and resistance to wear well. The last group of an alloy is Group C, which comprises LM12, LM26, LM28, LM29, LM30, and LM31. These alloys are used only for limited applications where high strength and shock resistance are needed, and the thermal expansion coefficient is weak. In the automotive industry Group, C alloys are primarily used to manufacture pistons [2].

Aluminium is extensively used in the automobile, aerospace industry, and other engineering applications because of its lightweight properties. The poor strength and low melting point of aluminium is a significant problem for aluminum-based products [3,4]. To avoid these kinds of problems, ceramic particles with aluminum can be added as an aluminum matrix composite, resulting in new material with a lightweight and higher strength [5]. A composite is a substance composed of two or more physically and chemically distinct parts that are properly assembled to obtain a new material with unique properties compared with the base composition. Several methods are used to produce composites of aluminium matrix that are solid-state processes, liquid state processes, deposition, and *in situ* processes [6].

Metal matrix composites have superior strength and stiffness at elevated temperatures and are used where better mechanical properties are needed. The reinforcing process encourages applications that take advantage of the physical properties of reinforcement and exhibit the entire properties of composites. Reinforcement materials can be categorized as phases of discontinuous strengthening and continuous fiber reinforcement. In general, discontinuous reinforcements are categorized as particulate, short, or chopped fiber and whisker. A few examples are silicon carbide, alumina or carbon, *etc.* [7]. Particles of reinforcement used to create aluminum composites are SiC, Al_2O_3, Glass Fibre, TiO_2, and B_4C, *etc.* These particles aid in improving high strength to achieve better mechanical and tribological properties [3]. Aluminium matrix composite is broadly used because of good mechanical properties such as high strength to weight ratio high wear resistance, compared to other materials [8]. The focus of this review is to examine composite production methods based on matrix material, mechanical properties, and tribological behavior.

METHODS OF PREPARING COMPOSITES

Assorted methods such as a solid-state process, liquid state process and deposition process are required to manufacture aluminum composite material. Diffusion bonding and physical vapor deposition methods are practiced in making composite material in solid-state process power blending and consolidation. Ins liquid state process, available methods are stir casting, compo casting, squeeze casting, spray forming, and liquid melt infiltration methods. The last available method is the deposition process. In this process, a particle is continuously moved through a region of metal to be deposited with high pressure. The deposition method has two additional forms: deposition by physical vapor and deposition by spray. Due to low manufacturing costs, the liquid state process is used to produce metal matrix composite. Due to its simplification and better results, the stir casting method is widely used among various methods of the liquid state process [6].

Stir Casting Process

Stir casting is a liquid state process in which two stages are present; the first is making molten metal, and the second is the introduction of reinforcement particles by *ex situ* or *in situ* method. Finally, the molten metal is poured into the preheated die to make a composite material. The process parameters are the key factors to produce quality and good composite parts.

Fabrication of Stir Casting Setup

For the fabrication of stir casting setup refractory stone, the crucible, blower, and stirrer are used as a component for construction. Refractory brick is used during the melting process to prevent heat losses. A crucible serves as a container where base metals and reinforcements are used for the melting process. A stirrer is an essential component of the stir casting system used to properly mix the molten metal, making the reinforcement particle uniform. Stir casting is commonly used for producing composite materials [9]. Two dissimilar methods are employed in the stir casting technique to manufacture composite material, such as the traditional and ultrasonic casting processes. In the ultrasonic system, reinforcement particles are homogeneously dispersed with increasing mechanical and physical properties [10]. While using this technique, we should meet a series of technical challenges to get good composite properties that include uniform distribution of reinforcement particles, wettability, porosity, and chemical reactions between matrix and reinforcement particles [11,12].

The difficulties of the stir casting system are achieving a homogeneous dispersion of ceramic particles introduced during the process and minimizing porosity in the composite material [13]. The operating parameters are maintained according to the process specifications to attain uniform distribution and high mechanical properties. The common operating parameters are stirrer speed, stirrer blade angle, number of blades, stirrer time, preheating temperature and weight percentage of reinforcement, die material, and preheating temperature [14].

Effect of Reinforcements in Composites

The addition of reinforcement particles is categorized into two approaches: *ex-situ* and *in-situ*. When the particles are introduced to the melt from outside, it is called the *ex-situ* process, and the same particle is synthesized within the melt is called the *in-situ* process [15]. In conventional methods, particles are mixed with the matrix using the *ex-situ* technique. This method is inefficient, and it builds non-uniform distribution of particles inside the matrix and produces poor wettability. These problems can be avoided by introducing *in-situ* techniques [16]. *In-situ* manufacturing involves exothermal reaction between ingredient and composites with aluminium molt. The key reward of this technique is good wettability between the particles and matrix, homogeneous distribution, fine particle size, and clean interface [17]. In the *ex-situ* technique, reinforcement particles are prepared separately and later added with the matrix. The *ex-situ* process is only controlled during the introduction of reinforcement particles, while in the *in-situ* process, it is controlled by respective process parameters. *In situ* method produces thermodynamically stable, high-temperature performance and strong bonding between reinforcement and matrix [18]. The mechanical and physical interaction between particle and matrix can be enhanced in *in-situ* technique because of its uniform particle distribution [19].

Manufacturing of Aluminium Based Metal Matrix Composites

Composites made of an aluminium metal matrix are prepared using a stir casting setup. Initially, the crucible's inner surface is thoroughly washed to remove dirt. The fresh aluminum matrix is heated above the furnace melting temperature. To extract the moisture content to be combined with the aluminum melt or reinforcement particles are directly poured into the aluminum melt, reinforcement particles are heated separately. The stirring process would be done during the addition of hard particles to disperse the particles evenly to ensure consistent composite melting properties. To get composite metal shaped, molten metal is poured into the preheated die. The properties of the cast are based on the process parameters of the technique [20].

Review on Mechanical and Wear Properties of Composites Based on Matrix

Aluminium Alloy

The mechanical, thermal, and tribological properties of aluminium-based alloy composites are improved by various reinforcements like SiC, Al_2O_3, B_4C, TiC, and $ZrSiO_4$, fly ash, remains of bauxite, groundnut shell, rice husk ash, and bagasse [21]. A higher weight percentage of silicon carbide will improve matrix hardness and tensile strength [1]. Aluminum cast composite is made from the cheapest strengthened particles such as alumina, Silicon carbide, graphite, fly ash, *etc*. Additional alumina to aluminum improves mechanical and tribological properties by applying coconut ash and fly ash to the part, improving its tensile strength [8]. Stirring of super-heated Al-melt elevates the melting temperature with the constant spreading of $Al_{13}Fe_4$ reinforcement particles in the Al matrix. Stirring may produce high hardness and stress without losing ductility [22]. Micron-size SiC particles are used in the aluminum matrix to produce Al-3Wt% SiC composite at two different temperatures and stirring times. Also, the composite has a higher ceramic connection when manufactured at a higher stirring temperature [23]. The tribological behavior of Al/SiC/Gr has been reviewed and concluded that the addition of reinforcement particles could enhance the wear performance of the composite, and the tribological properties were gradually decreased when adding Gr beyond the optimum value [24]. The multi-wall carbon nanotube is also used to increase the aluminum electrical resistivity [25].

Al6061 Alloy

The introduction of SiC as reinforcing particles in aluminum 6061 alloy improves the properties such as stiffness, tensile strength, and friction [26]. The mechanical properties such as hardness and compression are improved by applying 0 to 4 wt percentages Al_2O_3 particles [20]. The wear behavior was tested by incorporating TiB_2 as a weight ratio of 0, 4, 8 and 12, increasing wear resistance and mechanical properties by incorporating reinforcing particles [27]. The tribological behavior of Al6061/SiC/Ni-coated graphite (Gr-Ni) hybrid composite activity was studied to achieve maximum performance. The friction coefficient of a composite has been reduced by increasing the percentage of Gr-Ni. The optimum wear rate can be obtained when adding up to 5 percentages of Gr-Ni and it goes down when the Gr-Ni percentage crosses 8 percent [28]. The mixture of Cerium oxide and SiC/Al_2O_3 in the weight percentage of 0.5 to 2.5 and 2.5 to 7.5 was used as reinforcement. The inclusion of cerium oxide helps to refine the grain matrix. A composite flexural strength has risen to 44.76% while rising cerium oxide particle. Using optimum cerium oxide results in better composite output at 2.5% with an increased wear rate of 87.28% [29].

Al7075 Alloy

Mechanical properties of the Al-7075 metal matrix have been improved by adding various reinforcements such as SiC, Al_2O_3, Gr, and TiO_2 to unusual proportions of weight compared to base alloys [30]. By incorporating 12% B_4C and 3% coconut shell fly ash of reinforcement, the composite hardness has been improved to 33% by improving tensile power of 66% for 9% B4C and 3% CSFA for 9Wt % [31]. The hardness and tensile strength of a composite can be improved as a result of adding tungsten carbide as a reinforcing particle in the weight percentage of 2.5, 5.0, 7.5, 10, and 12.5 [32]. Wear behavior of aluminium 7075-boron carbide-coconut shell fly ash reinforced hybrid metal matrix composites was studied for various weight percentages of 0, 3, 6, 9, and 12 B_4C with 3 wt% CSFA composite. The inclusion of strong particles strengthened the composite's wear resistance [33]. A composite's tribological activity has been examined by introducing a 2-10 percentage weight of titanium carbide as a reinforcement element. A composite wear rate has decreased while the weight percentage of the reinforcement particle has increased. The optimum value of wear rate can be achieved by having 8 percentages of particles with a matrix compared to other matrices and composites [34].

A356 Alloy

The properties have been developed to measure optimum hardness and tensile strength using the Taguchi method with reinforcement of multi-wall carbon nanotube. Optimum of 106.4HV hardness value and 277.0 MPa of ultimate tensile strength were obtained, also identified that the hardness (76.3%) and ultimate tensile strength (108%) has been increased [35]. By adding SiC and Flyash particles as reinforcement can fabricate a defect-free aluminium matrix to improve the hardness, tensile strength, toughness and fatigue strength. A356/15%SiC/5% fly ash has produced better effects [36]. Mechanical properties like yield strength, ultimate tensile strength, hardness, and wear properties are improved when adding SiC with A356 alloy compared with base alloy [37]. The Zirconia particle is introduced to explore the characteristics of the tribological and microstructural properties in the weight proportion of 3, 6, 9, and 12. Further addition of Zirconia particles increases the ductility and strength of a composite as well as the particle distribution of cast content [38].

A413 Alloy

FGM microstructure, hardness, and dry sliding wear behavior have been observed in A413 alloy and Al-26wt % Ni master alloy by changing the weight percentage of Ni refinement particles in 3, 6, and 9. It was observed that the composite wear rate was lower when the Ni content was increased relative to the A413 alloy. The existence of silicone particles helps to reduce the composite wear rate [39]. Hardness and tensile behavior were tested with reinforcement of borosilicate powder that improves the composite's mechanical properties by reducing the particle size with increasing the addition of reinforcement [40]. The optimum process parameter of machinability has been found to achieve a maximum material removal rate with minimum surface roughness. The composites were made with the combination of fly ash and B_4C as reinforcement particles [41]. Copper-coated steel wire mesh is reinforced to increase the mechanical properties by varying the orientation *viz.*, 0°, $45^{O,}$ and 90°. It improves matrix-reinforcement bonding and tensile strength when the angle of wire mesh increases [42]. When using 2.5 – 10 wt% of copper coated steel fiber as reinforcement, the mechanical properties of composite such as hardness and tensile strength were improved by increasing the wt % of fiber, and tensile strength increased up to 19% for 5wt % composite fiber. Adding further wt% of fiber continuously reduces the weight loss, friction coefficient, and wear rate [43]. Using zinc-coated steel fiber instead of copper coated for the same angle gives a continuous improvement in tensile strength up to 11% at 90° of steel wire, and the ductility decreases as a result of the introduction of steel wire mesh [44]. Increasing the distance between matrix and steel wire improves the hardness of the composite. Adding three steel wires improves the tensile strength by up to 24%. At the same time, increasing the number of wires embedded in the matrix reduces weight loss, wear rate, and coefficient of friction [45].

LM13 Alloy

The use of coated steel fibers as reinforcement eliminates weight losses in the composites by consistent reinforcing distribution in the structure [46]. The characterization of the microstructural, mechanical, and wear properties of aluminum alloy LM13 ($Al\text{-}Si1_2CulMg_2$) were investigated by incorporating titanium diboride and boron carbide refining particles using the stir casting process. With the increase in weight percentage of reinforcements, hardness and tensile strength have been improved, and the distribution of reinforcement is uniform. The loss of weight of the material also decreased as the load and sliding distance were increased [47]. Reinforced titanium carbide (TiC) and zirconium diboride (ZrB_2) particles were used to make a composite part to explore the impact of a sliding distance. The composite parts have higher mechanical characteristics than the non-enhanced aluminum alloy [48]. In contrast to traditional materials, the composite can provide high strength to weight ratio, specific strength, creep resistance, and low density [49].

Copper-coated steel wire mesh is reinforced to increase the mechanical properties and wear resistance. Tensile strength is increasing up to 28% with increasing up to three numbers of wire meshes. When the wire mesh increases to five, the weight loss falls to 69% [50], and the weight loss decreases as increasing wt% of reinforcement and stirring speed [51]. The hardness of a composite has been examined with the addition of 0-10wt% of reinforcing particles. Hardness is improved when increasing the wt% of reinforcement particle and stirrer speed [52]. Tensile strength increases with increasing wt% of reinforcement up to 6% also increases due to the uniform distribution while increasing stirrer level [53].

LM26 Alloy

A composite was tested with SiC as reinforcement, which increases the wear resistance by increasing the level of reinforcing particle in the weight percentage of 5, 10, and 15 [54]. The properties of a composite increased while the wt% of reinforcement particle increased as 0, 2, 4, 6, and 8 as well as the optimum wear resistance can be accomplished at 6wt% [55], the optimization of hardness and tensile strength was performed by using the Taguchi method [56]. The void fraction decreases up to 0.36% by increasing the wt% of porcelain ceramic particles up to 6wt% and adding 8wt%, which increases to 0.65%. After adding 8wt% of porcelain, microhardness increases up to 10%, and tensile strength increases by 20% for 6wt% of porcelain ceramic particles [57]. LM-based alloys are commonly used in automotive structures and components due to the low thermal expansion properties as well as fast casting capacity and cost-effective process. Usually, LM13 alloys are used for piston production. At present, LM26 is the best replacement material for automotive component manufacturing [58]. For this purpose, static structural analysis was performed using an Ansys tool to investigate the stress produced by the piston and piston ring [59]. An appropriate material has been selected for a piston with necessary properties by varying the wt% of recycled porcelain reinforcement particles by 0-8wt%. The best property needed for piston material is obtained at 6wt% of porcelain particle [60]. The findings of different reinforcement particles added with dissimilar matrix material with their respective output performance were clearly stated (Table **1**).

Table 1. Mechanical properties of Aluminium based metal matrix composites.

Base Alloy	Reinforcement Particle	Reinforcement wt%	Hardness	Tensile Strength	References
Al5083	Ferrotitanium (FeTi), Silicon carbide	30gTiFe + (2, 5 and 7wt %)	117.83HV	-	[1]
Al5083	Sic	3,5,8 and 10wt%	90BHN	260Mpa	[10]
LM25	TiB_2	10wt%	91HV	-	[15]
AA2219	TiB_2	5 and 10wt%	96HV	234Mpa	[16]
Al	Sic	1-4wt%	33BHN	95Mpa	[18]
AA7010	TiB_2	5, 7.5 and 10wt%	110.93HV	193.9Mpa	[19]
Al6061	Al_2O_3	0-4wt%	81HV	164Mpa	[20]
Aluminium	$Al_{13}Fe_4$	Al-0.08Fe-0.06Si and 2 wt.% Mg	82.7 ±2.3 VHN	185.2 Mpa	[22]
Aluminium	Micron- sized SiC	3wt%	90-100HV	120-140Mpa	[23]
Aluminium	multiwalled carbon nanotube	0.5, 1.0, 1.5, 2.0, 2.5 and 3.0 wt%	75.5HV	-	[25]
Al 6061	SiC	0, 2 and 4wt%	86VHN	157Mpa	[26]
Al 6061	TiB_2	0, 4, 8 and 12wt%	72.46VHN	137.86Mpa	[27]
Al6061	Cerium oxide and SiC/Al_2O_3	0.5-2.5wt% and 2.5 to 7.5wt%	92.8HV	123Mpa	[29]
Al7075	boron carbide and coconut shell fly ash	0, 3, 6, 9, 12wt% and 3wt%	169.5BHN	189Mpa	[31]
Al7075	tungsten carbide	2.5, 5.0, 7.5, 10 and 12.5wt%	97.2HV	152Mpa	[32]
AA7075	Titanium Carbide	2-10wt%	202.1HV	-	[34]
A356	multiwalled carbon nanotube	0.5-1.0wt%	106.4HV	277Mpa	[35]
A356	SiC and flyash	15wt% and 5wt%	97.65BHN	364.55Mpa	[36]
A356	Zirconia	3, 6, 9 and 12wt%	75.36BHN	200.22Mpa	[38]
A413	Ni	3, 6 and 9wt%	85BHN	-	[39]
A413	Borosilicate powder	0, 2.5, 5, 7.5 and 10wt%	84.1HV	174.18Mpa	[40]
A413	Copper coated steel wire mesh	0°, 45° and 90°	161HV	164Mpa	[42]
A413	Copper Coated Short Steel Fibers	2.5-10wt%	72BHN	173Mpa	[43]
A413	Zinc coated steel wire mesh	0°, 45° and 90°	96VHN	164Mpa	[44]

(Table 1) cont.....

Base Alloy	Reinforcement Particle	Reinforcement wt%	Hardness	Tensile Strength	References
A413	Zinc coated steel wire	1-5 Nos	611VHN	183Mpa	[45]
LM13	titanium diboride and boron carbide	0-10 and 5wt%	126BHN	174Mpa	[47]
LM13	zirconium diboride and titanium carbide	0, 2.5, 5, 10wt% and 5wt%	114BHN	238Mpa	[48]
LM13	Copper coated steel wire mesh	1-5 Nos	96BHN	138Mpa	[50]
LM13	Copper coated steel wire mesh	0-10wt%	128BHN	214Mpa	[52, 53]
LM 26	porcelain	0, 2, 4, 6, 8wt%	112.26HV	181.03Mpa	[55]
LM 26	rice husk ash and red mud	5, 10, 15 and 5wt%	82HV	220Mpa	[56]
LM26	porcelain ceramic particles	2, 4, 6 and 8wt%	112HV	181.03Mpa	[57, 60]

Applications

Aluminium alloy-based composites are frequently utilized in engineering applications that need lightweight, high stiffness, and moderate strength as indispensable properties. Because of their high specific strength and specific stiffness, aluminium alloy-based composites show high hardness and wear resistance, high thermal conductivity, high energy absorption, damping capacity, and low coefficients of friction and thermal expansion. The composite material can be tailored for automotive, aerospace, aircraft industries, defence, electronic packaging, commercial and industrial markets. Aluminum alloy composite is utilized in the manufacturing of golf club shafts and heads, skating shoes, baseball shafts, horseshoes, and bicycle frames in recreation and sports. Valves, crankshafts, gear parts and suspension arms, automotive push rods, piston and cylinder liners are some of the specialized applications of aluminium alloy-based composites [61].

CONCLUSION

Aluminium-based composite alloy metal is manufactured to meet the engineering requirements in the aerospace, shipbuilding, and automobile sectors because of its lightweight and high mechanical properties. Due to their simple manufacturing methods and low manufacturing costs, the liquid state process technique is widely used for the manufacturing of composite material. Metal matrix composites based on aluminum will increase the mechanical, physical, and tribological properties by adding reinforcing particles during the casting process. The inclusion of reinforcing particles is chosen based on the applications of composite material. The properties and characteristics of the composites are varied, depending on the properties of reinforcement particles and operating parameters of the casting process. Highly effective and specialized techniques are used to assess the properties and characteristics of the composite material.

CONSENT FOR PUBLICATION

Not applicable.

CONFLICT OF INTEREST

The author declares no conflict of interest, financial or otherwise.

ACKNOWLEDGEMENTS

Declared none.

REFERENCES

[1] S.O. Akinwamide, S.M. Lemika, B.A. Obadele, O.J. Akinribide, B.T. Abe, and P.A. Olubambi, "A Study on Microstructural and Mechanical Properties of a Stir Cast Al (SiC-Mg-TiFe) Composite", *Fluid Dynamics & Materials Processing,* vol. 15, no. 1, pp. 15-26, 2019. [http://dx.doi.org/10.32604/fdmp.2019.04761]

[2] Standard, British. , "Specification for Aluminium and Aluminium Alloy Ingots and Castings for General Engineering Purposes", *British Standard Institution,* 1988.

[3] G. Singh, and A. Kumar, "A Review Paper on Stir Casting of Reinforced Aluminium Metal Matrix Composite (MMC)", *International Journal of Current Engineering and Technology,* vol. 9, no. 3, pp. 432-439, 2019. [http://dx.doi.org/10.14741/ijcet/v.9.3.14]

[4] K.L. Chaitanya, and K. Srinivas, "A Review on Use of LM Based Aluminium Alloys in Automobile Components", *American International Journal of Research in Science, Technology, Engineering & Mathematics,* vol. 24, no. 1, pp. 103-107, 2018.

[5] M. Kumar, R.K. Gupta, and A. Pandey, "A Review on Fabrication and Characteristics of Metal Matrix Composites Fabricated by Stir Casting", *International Conference on Mechanical, Materials and Renewable Energy,* 2018. [http://dx.doi.org/10.1088/1757-899X/377/1/012125]

[6] P. Garg, A. Jamwal, D. Kumar, K.K. Sadasivuni, C.M. Hussain, and P. Gupta, "Advance research progresses in aluminium matrix composites: manufacturing & applications", In: *J. Mat. Res.Technol.,* vol. 8. , 2019, no. 5, pp. 4924-4939.

[7] "Structural materials handbook -Part 5: New advanced materials, advanced metallic materials, general design aspects and load transfer and design of joints, ECSS-E-HB-32-20 Part 5A, 20",

[8] K.K.S. Himanshu Kala, "Mer, Sandeep Kumar, "A Review on Mechanical and Tribological Behaviors of Stir Cast Aluminum Matrix Composites", 3rd International Conference on Materials Processing and Characterization, Procedia", *Mater. Sci.,* vol. 6, pp. 1951-1960, 2014.

[9] B.T. Ramesh, "Vinayak Koppad, Hemanth Raju T, "Fabrication of Stir casting Setup for Metal Matrix Composite", *International Journal for Scientific Research & Development,* vol. 5, no. 06, pp. 944-949, 2017.

[10] A.H. Idrisi, and A-H.I. Mourad, "Conventional stir casting *versus* ultrasonic assisted stir casting process: Mechanical and physical characteristics of AMCs"., *J. Alloys Compd.*, vol. 805, pp. 502-508, 2019.
[http://dx.doi.org/10.1016/j.jallcom.2019.07.076]

[11] K. Rajendra, M.V.S. Babu, P. Govinda Rao, and K.N.S. Suman, "Recent Developments in the Fabrication of Metal Matrix Composites by stir Casting route – a review", *Int. J. Chem. Sci.*, vol. 14, no. 4, pp. 2358-2366, 2016.

[12] K. Almadhoni, and S. Khan, "“Review of Effective Parameters of Stir Casting Process on Metallurgical Properties of Ceramics Particulate Al Composites”, IOSR Journal of Mechanical and Civil Engineering, Vol.12, Issue 6", *Ver.*, vol. IV, pp. 22-40, 2015.

[13] P. Wagh, A. Paliwal, R. Rajput, P. Potdar, and A.P. Kulkarni, "Aluminum Metal Matrix Composite Fabrication Using Stir Casting: A Review", *Vishwakarma Journal of Engineering Research,* vol. 2, no. 4, pp. 159-164, 2018.

[14] M. Anguraj, and M. Easwaramoorthi, "Samson Jerold Samuel Chelladurai, “Mechanical Properties And Wear Behaviour of Aluminium Matrix Composites–An Overview", *Int. J. Sci. Adv. Res. Technol.*, vol. 3, no. 10, 2017.

[15] J. Dinesh, G. Mohammad Ashraf, and N. Radhika, "Fabrication and Characterization of Al LM25/Tib2 *in-situ* Composites"., *J. Eng. Appl. Sci. (Asian Res. Publ. Netw.),* vol. 11, no. 9, pp. 6001-6005, 2016.

[16] N.R. Rajasekaran, and V. Sampath, "Effect of *in-situ* TiB2 Particle Addition on the Mechanical Properties of AA 2219 Al Alloy Composite"., *J. Miner. Mater. Charact. Eng.*, vol. 10, no. 6, pp. 527-534, 2011.
[http://dx.doi.org/10.4236/jmmce.2011.106040]

[17] C.A. Lawrance, "Dr. P. Suresh Prabhu, “Combined Effects of Stir Casting and *in-situ* Techniques in The Production of Al 6061-TiB2 Metal Matrix Composite with Different Reaction Holding Times and Its Characterizatio"., *International Journal of Mechanical Engineering and Technology,* vol. 9, no. 3, pp. 62-70, 2018.

[18] M. Senemar, B. Niroumand, A. Maleki, and K. Pradeep, "Rohatgi, “Synthesis of the *in-situ* aluminum matrix composite through pyrolysis of high temperature vulcanization silicone"., *J. Compos. Mater.*, vol. 52, no. 1, pp. 123-134, 2018.
[http://dx.doi.org/10.1177/0021998317702955]

[19] N. Mallireddy, and K. Siva, "Investigation of Microstructural, Mechanical and Corrosion Properties of AA7010-TiB2 *in-situ* Metal Matrix Composite"., *Sci. Eng. Compos. Mater.*, vol. 27, no. 1, pp. 97-107, 2020.
[http://dx.doi.org/10.1515/secm-2020-0010]

[20] S. Sivananthan, V. Rajalaxman Reddy, and C. Samson Jerold Samuel, "Preparation and evaluation of mechanical properties of 6061Al-Al2O3 metal matrix composites by stir casting process", *Mater. Today Proc.*, vol. 21, pp. 713-716, 2020.
[http://dx.doi.org/10.1016/j.matpr.2019.06.744]

[21] S. Kant, and A.S. Verma, "Stir Casting Process in Particulate Aluminium Metal Matrix Composite: A Review", *International Journal of Mechanics and Solids,* vol. 12, no. 1, pp. 61-69, 2017.

[22] S.K. Pradhan, S. Chatterjee, A.B. Mallick, and D. Das, "A simple stir casting technique for the preparation of *in situ* Fe-aluminides reinforced Al-matrix composites"., *Perspect. Sci. (Neth.),* vol. 8, pp. 529-532, 2016.
[http://dx.doi.org/10.1016/j.pisc.2016.06.011]

[23] S. Soltani, R.A. Khosroshahi, R.T. Mousavian, Z-Y. Jiang, A.F. Boostani, and D. Brabazon, "Stir casting process for manufacture of Al–SiC composites", *Rare Met.*, vol. 36, no. 7, pp. 581-590, 2017.
[http://dx.doi.org/10.1007/s12598-015-0565-7]

[24] J. Singh, "Fabrication characteristics and tribological behavior of Al/SiC/Gr hybrid aluminum matrix

composites: A review", *Friction,* vol. 4, no. 3, pp. 191-207, 2016.
[http://dx.doi.org/10.1007/s40544-016-0116-8]

[25] S. Sivananthan, S. Gnanasekaran, and C. Samson Jerold Samuel, "Preparation and Characterization of Aluminium Nanocomposites Based on MWCNT", *Appl. Mech. Mater.,* vol. 550, pp. 30-38, 2014.
[http://dx.doi.org/10.4028/www.scientific.net/AMM.550.30]

[26] S. Sivananthan, K. Ravi, and C. Samson Jerold Samuel, "Effect of SiC particles reinforcement on mechanical properties of aluminium 6061 alloy processed using stir casting route", *Mater. Today Proc.,* vol. 21, pp. 968-970, 2020.
[http://dx.doi.org/10.1016/j.matpr.2019.09.068]

[27] S. Suresh, and N. Shenbaga Vinayaga Moorthi, "Process development in stir casting and investigation on microstructures and wear behavior of TiB2 on Al6061 MMC", *Procedia Eng.,* vol. 64, pp. 1183-1190, 2013.
[http://dx.doi.org/10.1016/j.proeng.2013.09.197]

[28] M.L. Ted Guo, and C-Y.A. Tsao, "Chi.-Y.A. Tsao, "Tribological behavior of aluminum/SiC/nickel-coated graphite hybrid composites", *Mater. Sci. Eng. A,* vol. 333, no. 1-2, pp. 134-145, 2002.
[http://dx.doi.org/10.1016/S0921-5093(01)01817-2]

[29] V.K. Sharma, V. Kumar, and R.S. Joshi, "Investigation of rare earth particulate on tribological and mechanical properties of Al-6061 alloy composites for aerospace application", In: *J. Mater. Res. Technol.,* vol. 8. , 2019, no. 4, pp. 3504-3516.
[http://dx.doi.org/10.1016/j.jmrt.2019.06.025]

[30] A.R. Mohammed Imran, "Characterization of Al-7075 metal matrix composites: a review", In: *J. Mater. Res. Technol.,* vol. 8. , 2019, no. 3, pp. 3347-3356.

[31] B. Subramaniam, B. Natarajan, B. Kaliyaperumal, S. Jerold, and S. Chelladurai, "Investigation on mechanical properties of aluminium 7075-boron carbide-coconut shell fly ash reinforced hybrid metal matrix composites", *China Foundry,* vol. 15, no. 6, pp. 449-456, 2018.
[http://dx.doi.org/10.1007/s41230-018-8105-3]

[32] S. Arun Prakash, "Shaikh Anis Abdul, Razzak Ajay Christan F, Logesh M, "Mechanical Characteristics of AA7075 reinforced with Tungsten carbide produced by stir casting", *Int. J. Pure Appl. Math.,* vol. 119, no. 15, pp. 2015-2029, 2018.

[33] Balasubramani Subramaniam, Balaji Natarajan, Balasubramanian Kaliyaperumal, and Jerold Samuel Chelladurai Samson, "Wear behaviour of aluminium 7075—boron carbide- coconut shell fly ash reinforced hybrid metal matrix composites", *Mater. Res. Express 6,* 2019.
[http://dx.doi.org/10.1088/2053-1591/ab4052]

[34] V. Ramakoteswara Rao, N. Ramanaiah, and M.M.M. Sarcar, "Tribological properties of Aluminium Metal Matrix Composites (AA7075 Reinforced with Titanium Carbide (TiC) Particles)", *International Journal of Advanced Science and Technology.,* vol. 88, pp. 13-26, 2016.
[http://dx.doi.org/10.14257/ijast.2016.88.02]

[35] H. Hanizam, M.S. Salleh, M.Z. Omar, and A.B. Sulong, "Mohd ShukorSalleh, Mohd ZaidiOmar, Abu BakarSulong, "Optimisation of mechanical stir casting parameters for fabrication of carbon nanotubes–aluminium alloy composite through Taguchi method", *J. Mater. Res. Technol.,* vol. 8, no. 2, pp. 2223-2231, 2019.
[http://dx.doi.org/10.1016/j.jmrt.2019.02.008]

[36] S.P. Dwivedi, S. Sharma, and R.K. Mishra, "Microstructure and mechanical behavior of A356/SiC/Fly-ash hybrid composites produced by electromagnetic stir casting", *J. Braz. Soc. Mech. Sci. Eng.,* 2014.
[http://dx.doi.org/10.1007/s40430-014-0138-y]

[37] K Sunitha, and K Gurusami, "Effect of SiC Particulate Reinforcement in A356 / LM25 Alloy Composites and in their Hybrids", *International Journal of Recent Technology and Engineering,* vol. 8, no. 2S11, pp. 3984-3988, 2019.

[38] Karthikeyan Govindan, "Mechanical Properties and Metallurgical Characterization of LM25/ZrO2 Composites Fabricated by Stir Casting Method", *Revista Matéria,* 2019.

[39] A Saiyathibrahim, R Subramanian, and C Samson Jerold Samuel, "Processing and properties evaluation of centrifugally cast *in-situ* functionally graded composites reinforced with Al3Ni and Si particles"., *Mater. Res. Express 6,* 2019.

[40] S. Karthikeyan, "S, "Mechanical Behaviour of A413 alloy Reinforced with Borosilicate Powder", *Int. J. Appl. Eng. Res.,* vol. 10, no. 19, pp. 14655-14660, 2015.

[41] J. Udaya Prakash, and T.V. Moorthy, "J.Milton Peter, "Experimental Investigation on Machinability of Aluminium Alloy (A413)/Flyash/B4C Hybrid Composites Using wire EDM", *Procedia Eng.,* vol. 64, pp. 1344-1353, 2013.
[http://dx.doi.org/10.1016/j.proeng.2013.09.216]

[42] J.S.C. Samson, R. Arthanari, K. Krishnamoorthy, K.S. Selvaraj, and P. Govindan, "Effect of Copper Coating and Reinforcement Orientation on Mechanical Properties of LM6 Aluminium Alloy Composites Reinforced with Steel Mesh by Squeeze Casting", *Trans. Indian Inst. Met.,* 2017.
[http://dx.doi.org/10.1007/s12666-017-1235-2]

[43] J.S.C. Samson, R. Arthanari, N. Nithyanandam, K. Rajendran, and K.K. Radhakrishnan, "Investigation of Mechanical Properties and Dry Sliding Wear Behaviour of Squeeze Cast LM6 Aluminium Alloy Reinforced with Copper Coated Short Steel Fibers", *Trans. Indian Inst. Met.,* 2017.
[http://dx.doi.org/10.1007/s12666-017-1258-8]

[44] J.S.C. Samson, R. Arthanari, K. Krishnamoorthy, K.S. Selvaraj, and P. Govindan, "Investigation of The Mechanical Properties of a Squeeze-Cast Lm6 Aluminium Alloy Reinforced with A Zinc-Coated Steel-Wire Mesh", *Mater. Technol.,* vol. 52, no. 2, pp. 125-131, 2018.

[45] J.S.C.A.R.A. Samson, "Investigation on Mechanical and Wear Properties of Zinc-Coated Steel Wires Reinforced LM6 Aluminium Alloy Composites by Squeeze Casting", *Surf. Rev. Lett.,* vol. 1850125, pp. 1-10, 2018.

[46] J.S.C. Samson, and R. Arthanari, "Effect of stir cast process parameters on wear behaviour of copper coated short steel fibers reinforced LM13 aluminium alloy composites", *Mater. Res. Express,* vol. 5, no. 6, pp. 1-22, 2018.

[47] P. Vadivel, V. Chinnusamy, and S.J.S. Chelladurai, "Experimental investigation on mechanical properties and wear characterization of Al-Si12CulMg1 aluminium alloy reinforced with titanium diboride and boron carbide particulate hybrid composites using stir casting process", *Materialwiss. Werkstofftech.,* vol. 51, no. 1, pp. 73-82, 2020.
[http://dx.doi.org/10.1002/mawe.201800166]

[48] A. Ramaswamy, A.V. Perumal, and J.S.C. Samson, Investigation on Mechanical Properties and Dry Sliding Wear Characterization of Stir Cast LM13 aluminium alloy-ZrB_2-TiC Particulate Hybrid Composites., *Mater. Res. Express,* vol. 6, no. 6, pp. 1-31, 2019.
[http://dx.doi.org/10.1088/2053-1591/ab0ef8]

[49] N. Idusuyi, and J.I. Olayinka, "Dry sliding wear characteristics of aluminium metal matrix composites: a brief overview", In: *J. Mater. Res. Technol.,* vol. 8. , 2019, no. 3, pp. 3338-3346.

[50] J.S.C. Samson, R. Arthanari, R. Selvarajan, S. Athanarsamy, S. Arumugam, and G. Veerakumar, "Investigation on Mechanical Properties and Wear Behaviour of Squeeze Cast LM13 Aluminium Alloy Reinforced with Copper Coated Steel Wires", *Z. Phys. Chem.,* vol. 232, no. 12, pp. 1787-1806, 2018.
[http://dx.doi.org/10.1515/zpch-2017-1093]

[51] J.S.C. Samson, and R. Arthanari, "Effect of stir cast process parameters on wear behavior of copper coated short steel fibers reinforced LM13 aluminium alloy composites", *Mater. Res. Express,* vol. 5, no. 6, 2018.

[52] S.J.S. Chelladurai, and R. Arthanari, "Prediction of hardness of stir cast LM13 aluminum alloy -copper

coated short steel fiber reinforced composites using response surface methodology", *Materialwiss. Werkstofftech.,* vol. 51, no. 2, pp. 221-229, 2020. [http://dx.doi.org/10.1002/mawe.201900068]

[53] J.S.C. Samson, R. Arthanari, and R. Selvarajan, "makrishnan Kanagaraj, Palanisamy Angappan, "Investigation on microstructure and tensile behaviour of stir cast LM13 aluminium alloy reinforced with copper coated short steel fibers using response surface methodology", *Trans. Indian Inst. Met.,* vol. 71, no. 9, pp. 2221-2230, 2018. [http://dx.doi.org/10.1007/s12666-018-1353-5]

[54] Y. Avdhoot, "Vadghule, Prof. Vidyadhar C. Kale, "Tribological evaluation of LM26 Aluminum Metal Matrix Composites", *International Research Journal of Engineering and Technology,* vol. 5, no. 06, pp. 2553-2558, 2018.

[55] S. Dev, A. Aherwar, and A. Patnaik, "Preliminary Evaluations on Development of Recycled Porcelain Reinforced LM-26/Al-Si10Cu3Mg1 Alloy for Piston Materials", *Silicon,* vol. 11, no. 3, pp. 1557-1573, 2019. [http://dx.doi.org/10.1007/s12633-018-9979-9]

[56] "A. Pavan kumar, B. Jyothu naik, C H. Venkata rao, sadineni rama rao, "Optimization of Casting Parameters for casting of AL/RHA/RM hybrid composites using taguchi method", *International Journal of Engineering Trends and Technology,* vol. 4, no. 8, pp. 3284-3288, 2013.

[57] S. Dev, and A. Aherwar, "Study of Physico-Mechanical Properties of Porcelain Filled Aluminium LM26 Alloy for Piston Material", *Proceedings of IRF International Conference,* 2016 New Delhi, India978-93-85832-29-1

[58] K.L. Chaitanya, and K. Srinivas, "A Review on Use of LM Based Aluminium Alloys in Automobile Components", *American International Journal of Research in Science, Technology, Engineering & Mathematicsm,* vol. 24, no. 1, pp. 103-107, 2018.

[59] L.R. Arun, and M.A. Vinoth, "Vinay Kumar Uppinal, "Development and Assessment of Piston by using Al-Si Hybrid Metal Matrix Composites Reinforced with SiC and Cenosphere Particulates", *Int. J. Eng. Res. Technol. (Ahmedabad),* vol. 3, no. 7, pp. 1285-1289, 2014.

[60] S. Dev, A. Aherwar, and A. Patnaik, "Material Selection for Automotive Piston Component Using Entropy-VIKOR Method", *Silicon,* vol. 12, no. 1, pp. 155-169, 2020. [http://dx.doi.org/10.1007/s12633-019-00110-y]

[61] F. Nturanabo, *Leonard Masu and John Baptist Kirabira, "Novel Applications of Aluminium Metal Matrix Composites, Aluminium Alloys and Composites.* Kavian Omar Cooke, IntechOpen, 2019. [http://dx.doi.org/10.5772/intechopen.86225]

CHAPTER 5

Experimental Analysis of Different Machining Parameters on Epoxy Granite Composite

S. Nallusamy[1,*] and **S. Sundar**[2]

[1] *Department of Adult, Continuing Education and Extension, Jadavpur University, Kolkata-700032, India*

[2] *Department of Mechanical Engineering, Dr. M.G.R. Educational and Research Institute, Chennai-600095, Tamilnadu, India*

Abstract: In the current manufacturing scenario, conventional machining processes play a broad role in manufacturing composite materials. In this research work, an end milling operation was carried out on reinforced epoxy granite composites. Experiments based on aggregate type, fiber content, spindle speed, and feed rate were performed using the Taguchi method. By undergoing a multi-level design, the L27 orthogonal array of Taguchi was obtained. Based on this design of experiment, an end milling operation was performed on prepared epoxy granite samples. The response constraints such as thrust force and surface roughness values were measured by appropriate measuring instruments. The outcome of system constraints was optimized based on response constraints. The optimization was performed through the multi-attribute decision-making technique for order preference by similarity to ideal solution. The ideal process constraints of spindle speed and feed rate were found 1200rpm and 60mm/min, respectively, which produce a better surface finish with minimal thrust force. Analysis of variance was used to verify the influence of process constraints on outcome response.

Keywords: ANOVA, End Milling, Epoxy Granite, Machining, Orthogonal Array, Polymer Composite, S/N Ratio, Spindle Speed, Surface Roughness, Taguchi Method, TOPSIS, Thrust Force.

INTRODUCTION

Epoxy Granite (EG) is a polymer matrix composite material where the granite and epoxy are in 80:20 compositions on a weight basis. The glass fiber and carbon fiber are added in 2% to the respective samples on a weight basis, and they are randomly orientated [1 - 3].

* **Corresponding author S. Nallusamy:** Department of Adult, Continuing Education and Extension, Jadavpur University, Kolkata-700032, India; E-mail: ksnallu@gmail.com

C. Samson Jerold Samuel, M. Suresh & S. Gnanasekaran (Eds.)

The machinability behavior of these composite under end milling processes is carried out, which provides cutting constraints for performing end milling operations on these samples. One of the statistical techniques of design of the experiment is used to determine the total number of experimental runs (N), which is obtained by the relation N=mk [4 - 7]. End milling operation was performed on fiber-reinforced plastic composite, and the result conveys that elevated cutting speed and lesser feed rate produce lower SR values [8 - 11]. End milling operation was performed on Inconel alloy optimization under Taguchi method, and gray relation analysis, which conveys that increase in material removal rate decreases SR while cutting velocity was found to be a critical constraint [12 - 15].

End milling operation was performed on aluminum alloy, and the result infers that cutting speed and depth of cut values affect the Surface Roughness (SR) [16 - 18]. The end milling process was executed on AISI I040 steel by minimizing SR and tool wear as an objective under regression analysis. As a result, the SR depends on spindle speed and feed rate, while the tool wear rate depends on machining time and depth of cut [19 - 21]. An end milling operation is performed on titanium alloy, which suggests that the cutting speed influences the SR that increases the cutting speed and decreases the SR so that a better surface finish is obtained [22 - 24]. In this research work, the samples are fabricated with varying epoxy-granite ratios on a weight basis, with and without fiber content. The end milling operation is carried out on these samples to produce a slot using a high-speed steel cutting tool in CNC vertical milling machine, and the SR values for the machined surface are measured. Based on the outcome response like thrust force and SR values, the input constraints are optimized using Technique for Order Preference by Similarity to Ideal Solution (TOPSIS) and validated with Analysis of Variance (ANOVA).

MATERIALS AND METHODS

The granite particle is used as a base material, and an epoxy resin is used as a polymer matrix to fabricate EG particle composites. The granite particles are obtained from crushing and sieving the black granite slabs. Araldite LY 556 is an epoxy resin used along with Aradur HY 951 hardener. The mixing ratio of hardener to resin is 1:10. The selected granite material and its size are indicated (Table **1**). The samples are fabricated for the composition of 80:20 granule mixtures. The sample of size of 125x125x10mm rectangular slab is selected for this study and obtained by ASTM standards. Type 4 aggregate is the mixture of 50% Type 1, 25% Type 2, and 25% Type 3 on a weight basis.

The following procedure is used for the preparation of EG composite samples.

Table 1. Aggregate type with particle size.

Aggregate Type	Size (Microns)
Type 1	106-599
Type 2	600-2799
Type 3	2800-11199

a. The mass of granite particle required for a single sample is calculated using the equation

$$\rho = m/\mathrm{v} \quad \textbf{(1)}$$

Where m is the required mass and V is the size of the die, and is the density of epoxy.

a. From the mass 'm' value, 80% indicates granite particle, and 20% indicates epoxy resin on a weight basis.
b. Epoxy resin and hardener are mixed properly in the ratio of 1:10.
c. The epoxy-hardener ratio is stirred properly with granite particles and placed inside the die cavity mounted on the shaker table.
d. After the die is filled with material, the die is un-mounted from the shaker table, and it is sealed properly with its top plate.
e. It is allowed to solidify for 24 hours, and the composite is removed from the die.
f. After removing the EG composite material from the die, it is placed inside the hot air oven for 6 to 10 hours to remove the further moisture content.
g. The glass fiber and carbon fiber are added to the required samples by 2% of their weight based on the fiber necessity.
h. The final prepared composite material (Fig. **1**) is then ready for operation.

Before performing an end milling operation on composite samples, the number of samples to be machined is determined from the design of the experiment through the Taguchi method. The four constraints are aggregate type, fiber content, spindle speed, and feed rate, and each constraint has 3 levels. The spindle speed and feed rate were chosen for machining operation are listed (Table **2**).

Fig. (1). EG composite sample.

Table 2. Cutting constraints.

Spindle Speed (rpm)	Feed Rate (mm/min)
600	30
1200	60
1800	90

With the above cutting constraints as input along with the aggregate type and fiber content, the Taguchi method is performed on these variables to obtain an orthogonal array as listed (Table **3**). The aggregate type levels are Type 1, Type 2, Type 4, and the fiber content levels are without fiber, glass fiber, and carbon fiber. Therefore, four factors with three levels create 27 experiment readings and are indicated from L1 to L27. By changing the spindle speed and feed rate for different aggregates and their corresponding outcome response will get varied, and hence, 27 samples have been fabricated.

Table 3. L27 orthogonal array by Taguchi method.

L Series	Aggregate	Fiber Content	Spindle Speed (rpm)	Feed Rate (mm/min)
L1	Type1	Without	600	30
L2	Type1	Without	600	30
L3	Type1	Without	600	30

(Table 3) cont.....

L Series	Aggregate	Fiber Content	Spindle Speed (rpm)	Feed Rate (mm/min)
L4	Type1	Without	1200	60
L5	Type1	Without	1200	60
L6	Type1	Without	1200	60
L7	Type1	Without	1800	90
L8	Type1	Without	1800	90
L9	Type1	Without	1800	90
L10	Type2	Glass	600	30
L11	Type2	Glass	600	30
L12	Type2	Glass	600	30
L13	Type2	Glass	1200	60
L14	Type2	Glass	1200	60
L15	Type2	Glass	1200	60
L16	Type2	Glass	1800	90
L17	Type2	Glass	1800	90
L18	Type2	Glass	1800	90
L19	Type4	Carbon	600	30
L20	Type4	Carbon	600	30
L21	Type4	Carbon	600	30
L22	Type4	Carbon	1200	60
L23	Type4	Carbon	1200	60
L24	Type4	Carbon	1200	60
L25	Type4	Carbon	1800	90
L26	Type4	Carbon	1800	90
L27	Type4	Carbon	1800	90

Process of End Milling

The process of end milling is carried out using BMV 40T20 model CNC vertical milling machine. Its spindle speed range is from 60 to 6000rpm, and a 10mm high-speed steel end milling cutter is used to perform the machining operation. The sample is mounted on the fixture, and the machining operation is carried out. At the time of machining, the forces induced on the samples are detected using a milling dynamometer. Among the feed, tangential, and thrust forces, the thrust force values are considered for optimization, and the force values are listed (Table **4**). Thrust force will only be considered for optimization because from the literature, it is clear that the ideal cutting constraint must possess minimal thrust force values. Whenever the thrust force increases, the SR value will increase because of more vibration, so the surface finish (Fig. **2**) will decrease.

Table 4. Results of thrust force values related to end milling process.

L Series	Aggregate	Fiber Content	Spindle Speed (rpm)	Thrust Force (N)
L1	Type1	Without	600	40.27
L2	Type1	Without	600	32.72
L3	Type1	Without	600	43.71
L4	Type1	Without	1200	29.13
L5	Type1	Without	1200	47
L6	Type1	Without	1200	23.03
L7	Type1	Without	1800	35.16
L8	Type1	Without	1800	33.10
L9	Type1	Without	1800	21.28
L10	Type2	Glass	600	32.41
L11	Type2	Glass	600	29.36
L12	Type2	Glass	600	34.10
L13	Type2	Glass	1200	31.80
L14	Type2	Glass	1200	43.25
L15	Type2	Glass	1200	45.32
L16	Type2	Glass	1800	52.48
L17	Type2	Glass	1800	44.10
L18	Type2	Glass	1800	41.57
L19	Type4	Carbon	600	102.4
L20	Type4	Carbon	600	108.2
L21	Type4	Carbon	600	78.65
L22	Type4	Carbon	1200	65.60
L23	Type4	Carbon	1200	79.11
L24	Type4	Carbon	1200	95.20
L25	Type4	Carbon	1800	61.10
L26	Type4	Carbon	1800	73.23
L27	Type4	Carbon	1800	58.51

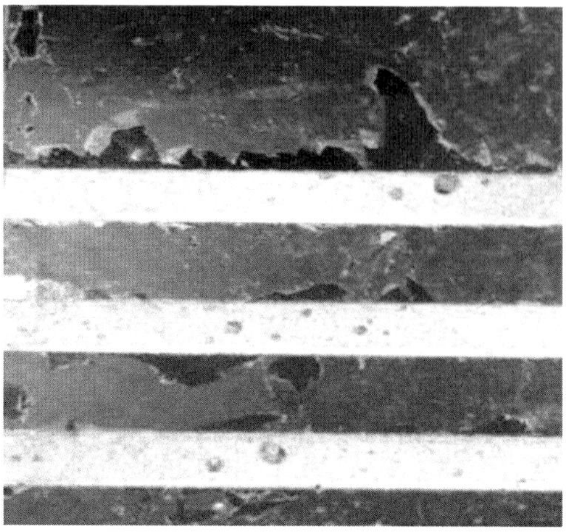

Fig. (2). Machined EG samples.

Measurement of Ra

The surface finish quality is incidental from the SR value (Ra) for the machined samples. The sample is placed over the flat table, and the SR meter probe is placed over the sample in three different places; its average value is considered as the SR values, and these values are tabulated (Table **5**).

Table 5. Results of RA for EG specimens.

L Series	Aggregate	Fiber Content	Surface Roughness (μm)
L1	Type1	Without	3.795
L2	Type1	Without	3.217
L3	Type1	Without	2.810
L4	Type1	Without	2.461
L5	Type1	Without	4.515
L6	Type1	Without	4.704
L7	Type1	Without	3.360
L8	Type1	Without	2.774
L9	Type1	Without	2.968
L10	Type2	Glass	7.880
L11	Type2	Glass	7.248
L12	Type2	Glass	2.737
L13	Type2	Glass	5.427
L14	Type2	Glass	4.573
L15	Type2	Glass	2.726

(Table 5) cont.....

L Series	Aggregate	Fiber Content	Surface Roughness (μm)
L16	Type2	Glass	2.614
L17	Type2	Glass	4.080
L18	Type2	Glass	3.330
L19	Type4	Carbon	2.924
L20	Type4	Carbon	3.171
L21	Type4	Carbon	3.141
L22	Type4	Carbon	3.110
L23	Type4	Carbon	3.702
L24	Type4	Carbon	2.913
L25	Type4	Carbon	2.220
L26	Type4	Carbon	3.648
L27	Type4	Carbon	2.802

Optimization Techniques

TOPSIS technique gives the result which is not only the closest to the hypothetically best and also farthest from the hypothetically worst. The different stages of solving by the TOPSIS technique are,

Stage 1: To find out the objective and recognize the parametric evaluations attribute.

Stage 2: To create a decision matrix Dm of m x n, where m is alternative and m attributes.

$$D_m = \begin{pmatrix} q_{11} & q_{12} & \cdots & q_{1n} \\ q_{21} & q_{22} & \cdots & q_{2n} \\ \vdots & & \ddots & \vdots \\ q_{m1} & q_{m2} & \cdots & q_{nm} \end{pmatrix} \quad \textbf{(2)}$$

Where q_{ij} is the effect of ith substitute related to jth attribute

Stage 3: To obtain the normalized matrix with the following term

$$r_{ij} = \frac{q_{ij}}{\sqrt{\sum_{i=1}^{m} q_{ij}^2}} \quad j = 1,2, \ldots\ldots n \quad \textbf{(3)}$$

Stage 4: Weight of every attribute was assumed to be w_j (j = 1, 2, . . ., n) and weighted normalized decision matrix U = [u_{ij}] can be obtained by

$$U = w_j \, r_{ij} \text{ where } \sum_{j=1}^{n} W_j = 1 \tag{4}$$

Stage 5: The positive and negative idyllic solutions were attained by

$$U^+ = \left\{ \left(\sum_i^{max} u_{ij} \,|j€J\right) \left(\left(\sum_i^{min} u_{ij}\right) \middle| (j €J|i = 1,2, \dots .m) \right) \right\} \tag{5}$$

$$U^+ = \{u_1^+, u_2^+, u_3^+, \dots .., u_m^+\} \tag{6}$$

$$U^+ = \left\{ \left(\sum_i^{min} u_{ij} \,|j€J\right) \left(\left(\sum_i^{max} u_{ij}\right) \middle| (j €J|i = 1,2, \dots .m) \right) \right\} \tag{7}$$

$$U^- = \{u_1^-, u_2^-, u_3^-, \dots .., u_m^-\} \tag{8}$$

Stage 6: Division among various alternatives were found by 'ideal' solution and expressed as

$$S_i^+ = \sqrt{\sum_{j=1}^{n} \left(u_{ij} - u_j^+\right)^2}, \quad i = 1,2, \dots, m \tag{9}$$

Division of alternatives from negative idyllic solution is expressed as

$$S_i^- = \sqrt{\sum_{j=1}^{n} \left(u_{ij} - u_j^-\right)^2}, \quad i = 1,2, \dots, m \tag{10}$$

Stage 7: Relative nearness of distinctive alternative to the positive ideal result is intended by

$$P_i = \frac{S_i^-}{S_i^+ + S_i^-} \quad i = 1,2, \dots .., m \tag{11}$$

Stage 8: To identify the set of alternatives having most and least preferred solutions, the P_i value was ranked in descending order. The preferred values must possess minimum thrust force with minimum SR, and based on that, the corresponding P_i values are listed (Table **6**) along with their rank by using the TOPSIS procedure from step 1 to step 8.

Table 6. Preference value assessment with rank order.

L Series	Performance Value (P_i)	Rank Order
L1	0.752	10
L2	0.846	5
L3	0.804	7
L4	0.930	1
L5	0.647	17
L6	0.724	12
L7	0.821	6
L8	0.880	3
L9	0.908	2
L10	0.476	25
L11	0.518	23
L12	00.875	4
L13	0.634	19
L14	0.668	16
L15	0.793	8
L16	0.751	11
L17	0.703	13
L18	0.782	9
L19	0.466	26
L20	0.436	27
L21	0.558	20
L22	0.637	18
L23	0.521	22
L24	0.494	24
L25	0.702	14
L26	0.557	21
L27	0.701	15

The table shows that L4 attains the first rank whose aggregate type is 1 with glass fiber at a spindle speed of 1200rpm with a feed rate of 60mm/min, which produces a thrust force of 29.13N and SR of 2.461µm. This is the best ideal solution, and these cutting constraints are chosen to carry out the end milling process. The second rank is attained by L9, whose arrogate type is 1 with carbon fiber at a spindle speed of 1800rpm with a feed rate of 90mm/min, whose thrust force value is 21.28N and the Ra is 2.968µm. This is considered as the next best ideal solution.

RESULT AND DISCUSSION

Results of Thrust Force

The minimum thrust force will produce minimum vibration, which leads to a better surface finish on the workpiece, so the objective is to select the cutting constraints, which produce minimum thrust force with minimum SR. At this end milling process, the minimum thrust force is 21.28 N, which was developed by a Type 1 sample of carbon fiber whose SR value is 2.968 µm, but the least SR value is 2.220µm developed on Type 4 carbon fiber sample. So the cutting constraints corresponding to minimum thrust force cannot be considered an ideal cutting condition.

Results of SR

To obtain a better surface finish, the value of SR should be minimum. At this end milling process, the minimum SR value is 2.220 µm, which was developed on a Type 4 sample of carbon fiber whose thrust force value is 61.10 N but the least thrust force value is 21.28N developed on Type 1 samples of carbon fiber. So the cutting constraint corresponding to the minimum SR value alone cannot be considered an ideal cutting condition because the ideal cutting condition must possess minimum SR with minimal thrust force values.

Ideal Cutting Constraints

The ideal cutting conditions will develop a better surface finish, which was indicated by minimum SR value with minimal thrust force. As a result of the TOPSIS multi-attribute decision-making method, the minimum SR value with minimal thrust force was obtained on a Type 1 sample of glass fiber whose cutting conditions are 1200rpm of spindle speed of 0.62m/s or 37.69m/min of cutting speed with the feed rate of 60mm/min. This ideal cutting condition will develop 29.13N thrust force with 2.461µm of SR on the EG composite, indicating that it produces a better surface finish than other cutting conditions under machining using a 10mm high-speed steel end milling cutter.

RESULTS OF ANOVA

Investigation of S/N Ratio

The objective of the S/N ratio analysis is to obtain minimum SR with minimum thrust force. So in S/N ratio analysis 'smaller is better' characteristic is chosen and it is obtained by using the relation as given below

$$\frac{S}{N} = -10 \log \frac{1}{n} (\sum y^2) \qquad \textbf{(12)}$$

The above relation is used for both SR (Fig. **3**) and thrust force (Fig. **4**) with S/N ratio analysis is plotted.

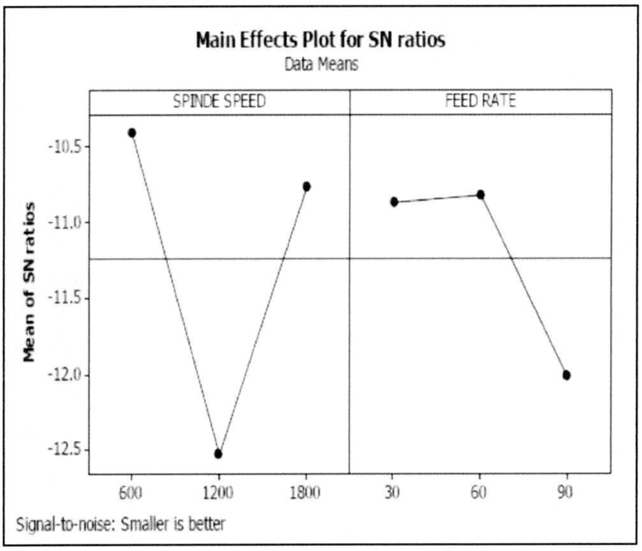

Fig. (3). S/N ratio analysis for SR.

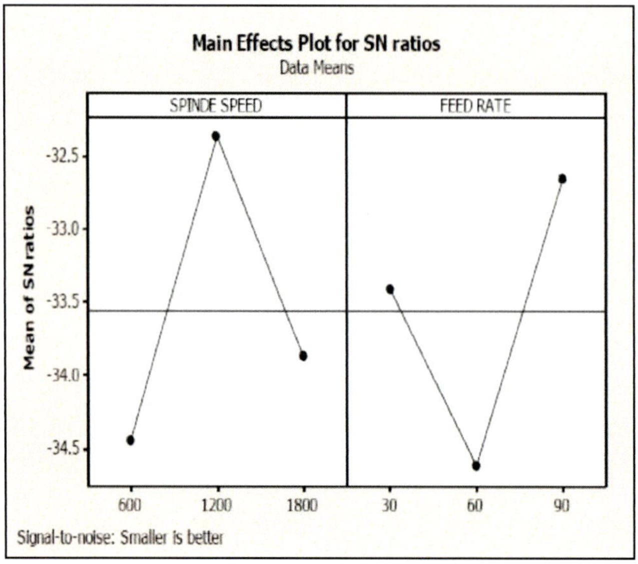

Fig. (4). S/N ratio analysis for thrust force.

From the S/N ratio analysis for SR plot, it is clear that, at lower spindle speed, S/N ratio value is maximum, and for a higher feed rate, S/N ratio value is minimum for SR. From the S/N ratio analysis for the thrust force plot, it is clear that the S/N ratio value is minimum for lower spindle speed, and the value of S/N ratio is maximum for a higher feed rate for thrust force.

Investigation of Thrust Force

The authority of spindle speed and feed rate on thrust force obtained from analysis of variance are tabulated (Table **7**). The table shows that spindle speed contributes 54.8% for thrust force and feed rate contributes 45.16% for thrust force, and its mean distribution values are placed (Fig. **5**). From the plot, it is clear that to obtain a minimum thrust force; the spindle speed should be maintained at 1200rpm.

Table 7. Results of two way ANOVA for thrust force.

Source	Summing Square (SS)	P Factor	% of Contribution
Spindle Speed	883.6	0.023	54.7
Feed Rate	727.6	0.041	45.15
Total	1611.2	-	-

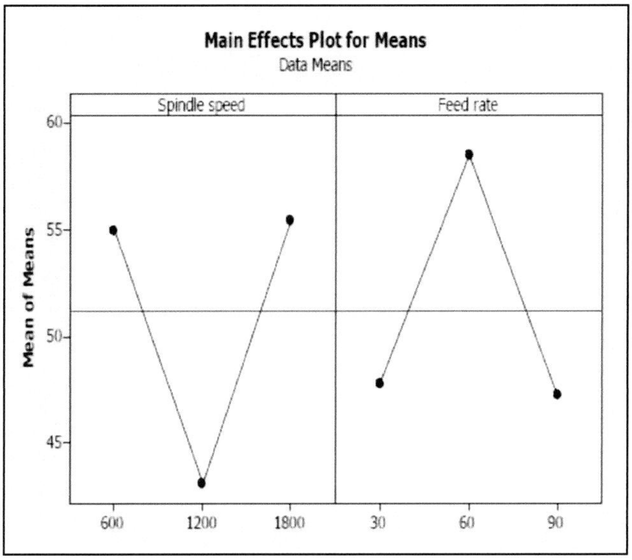

Fig. (5). Thrust force mean value Vs spindle speed and feed rate.

SR Investigation

The influence of spindle speed and feed rate on SR obtained from analysis of variance are tabulated (Table **8**).

Table 8. Results of two way ANOVA for SR.

Source	Summing Square (SS)	P Factor	% of Contribution
Spindle Speed	4.7468	0.218	66.98
Feed Rate	2.3385	0.457	33
Total	7.0853		

The table shows that spindle speed contributes 66.98% for SR and feed rate contributes 33% for SR, and its mean distribution is placed (Fig. **6**). From the plot, it is clear that to obtain a minimum SR, the feed rate should be maintained as 60mm/min.

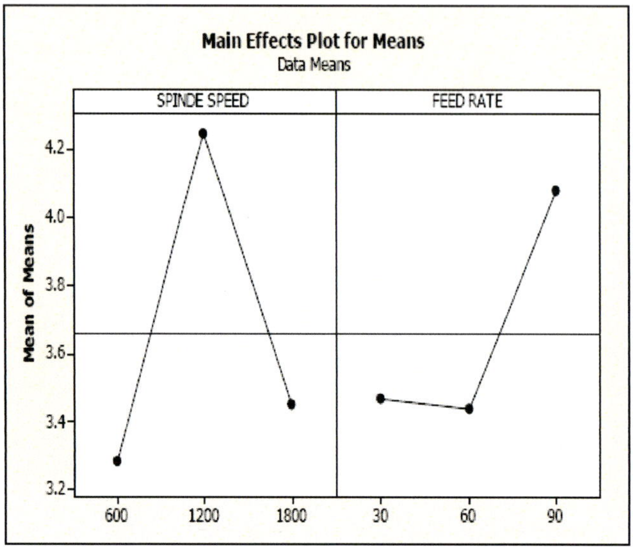

Fig. (6). SR mean value *Vs* spindle speed and feed rate.

CONCLUSION

In this research work, the L27 orthogonal array was designed using the Taguchi method for fabricated EG samples machined by 10mm end milling cutter. The ideal cutting constraints based on thrust force and SR were determined using TOPSIS, and the variance analysis was performed. Based on these evaluations, the following conclusions were made:

- The end milling process was performed on the EG samples as per the design of the experiment, and the thrust force progressed, and SR was measured.
- The minimum thrust force progressed was 21.28N with 2.968µm of SR for the Type 1 sample machined at 1800rpm spindle speed and 90mm/min feed rate.
- The minimum SR value was 2.220µm with 61.10N thrust force progressed on the Type 4 sample machined at 1200rpm spindle speed and 30mm/min feed rate.
- The minimum SR value doesn't contain minimum thrust force and *vice versa*. The TOPSIS multi-attribute decision-making method was used to determine the ideal cutting constraint to eliminate this contradiction.
- As a result of the TOPSIS method, the ideal cutting constraint obtained was 1200rpm of spindle speed, and 60mm/min of feed rate, which develops the minimal thrust force of 29.13N on the Type 1 glass fiber sample and its corresponding SR is 2.461µm.

- ANOVA results show that the contribution of spindle speed to obtain minimal thrust force and minimal SR was higher than the feed rate, and hence, spindle speed contributes more than the feed rate.

CONSENT FOR PUBLICATION

Not applicable.

CONFLICT OF INTEREST

The author declares no conflict of interest, financial or otherwise.

ACKNOWLEDGEMENTS

Declared none.

REFERENCES

[1] A. Joshi, "Pradeep Kothiyal. Investigating the effect of machining parameters of CNC milling on surface finish by Taguchi method", *International Journal on Theoretical and Applied Research in Mechanical Engineering,* vol. 2, no. 2, pp. 113-118, 2013.

[2] S.S.R. Raj, J.E.R. Dhas, and C. Jesuthanam, "Challenges on machining characteristics of natural fiber-reinforced composites – A review", *J. Reinf. Plast. Compos.,* vol. 40, no. 1-2, pp. 41-69, 2021. [http://dx.doi.org/10.1177/0731684420940773]

[3] O.T. Adesina, T. Jamiru, E.R. Sadiku, O.F. Ogunbiyi, and L.W. Beneke, "Mechanical evaluation of hybrid natural fibre-reinforced polymeric composites for automotive bumper beam: A review", *Int. J. Adv. Manuf. Technol.,* vol. 103, no. 5-8, pp. 1781-1797, 2019. [http://dx.doi.org/10.1007/s00170-019-03638-w]

[4] J. Eduardo, R. Manuelraz, and C.H. Lopes, "Optimization of machining parameters to improve the surface quality", *Procedia Struct. Integr.,* vol. 5, pp. 355-362, 2017. [http://dx.doi.org/10.1016/j.prostr.2017.07.182]

[5] B. Sidda Reddy, J. Suresh Kumar, and K. Vijaya Kumar Reddy, "Optimization of surface roughness in CNC end milling using response surface methodology and genetic algorithm", *Int. J. Eng. Sci. Technol.,* vol. 3, no. 8, pp. 102-109, 2011. [http://dx.doi.org/10.4314/ijest.v3i8.8]

[6] A.K. Sahoo, A.N. Baral, A.K. Rout, and B.C. Routra, "Multi-objective optimization and predictive modelling of surface roughness and material removal rate in turning using grey relational and regression analysis", *Procedia Eng.,* vol. 38, pp. 1606-1627, 2012. [http://dx.doi.org/10.1016/j.proeng.2012.06.197]

[7] B. Kursuncu, and Y.E. Biyik, "Optimization of cutting parameters with Taguchi and grey relational analysis methods in MQL-assisted face milling of AISI O2 steel", *J. Cent. South Univ.,* vol. 28, no. 1, pp. 112-125, 2021. [http://dx.doi.org/10.1007/s11771-021-4590-4]

[8] S. Chandrabakty, I. Renreng, Z. Djafar, and H. Arsyad, "An optimization of the machining parameters on delamination in drilling ramie woven reinforced composites using Taguchi method", *J. Phys. Conf. Ser.,* vol. 1341, no. 5, pp. 1-16, 2019. [http://dx.doi.org/10.1088/1742-6596/1341/5/052005]

[9] V.K. Vankanti, and V. Ganta, "Optimization of process parameters in drilling of GFRP composite using Taguchi method", *J. Mater. Res. Technol.,* vol. 3, no. 1, pp. 35-41, 2014.

[http://dx.doi.org/10.1016/j.jmrt.2013.10.007]

[10] M. Al Hazza, "Surface roughness modeling in high speed hard turning using regression analysis", *Int. Rev. Mech. Eng.,* vol. 8, no. 2, pp. 431-436, 2014.

[11] E.Y.T. Adesta, M.H.F. Al Hazza, M.Y. Suprianto, and M. Riza, "Predicting surface roughness with respect to process parameters using regression analysis models in end milling", *Adv. Mat. Res.,* vol. 576, pp. 99-102, 2012.
[http://dx.doi.org/10.4028/www.scientific.net/AMR.576.99]

[12] R. Vinayagamoorthy, and M. Anthony Xavior, "Parametric optimization on multi-objective precision turning using grey relational analysis", *Procedia Eng.,* vol. 97, pp. 299-307, 2014.
[http://dx.doi.org/10.1016/j.proeng.2014.12.253]

[13] "Investigation on precision turning of titanium alloys", *Adv. Mat. Res.,* vol. 622-623, pp. 399-403, 2013.

[14] P.K. Kharwar, and R.K. Verma, "Multiple machining performance optimization in drilling of MWCNT/epoxy nanocomposites using TOPSIS and grey theory", *J. Adv. Manuf. Syst.,* vol. 20, no. 2, pp. 249-271, 2021.
[http://dx.doi.org/10.1142/S021968672150013X]

[15] K.R. Kumar, and V.S. Sreebalaji, "Modeling and analysis on the influence of reinforcement particle size during EDM of aluminum (Al/3.25Cu/8.5Si)/fly ash composites", *J. Adv. Manuf. Syst.,* vol. 15, no. 4, pp. 189-207, 2016.
[http://dx.doi.org/10.1142/S0219686716500141]

[16] F.J. Pontes, A.P. Paiva, P.P. Balestrassi, J.R. Ferreira, and M.B. Silva, "Optimization of radial basis function neural network employed for prediction of surface roughness in hard turning process using Taguchi's orthogonal arrays", *Expert Syst. Appl.,* vol. 39, no. 9, pp. 7776-7787, 2012.
[http://dx.doi.org/10.1016/j.eswa.2012.01.058]

[17] S. Nallusamy, and A. Karthikeyan, "Analysis of wear resistance, cracks and hardness of metal matrix composites with SiC additives and Al2O3 as reinforcement", *Indian J. Sci. Technol.,* vol. 9, no. 35, pp. 1-9, 2016.
[http://dx.doi.org/10.17485/ijst/2016/v9i35/100149]

[18] M.H.F. Al Hazza, E.Y.T. Adesta, M. Riza, and M.Y. Suprianto, "Surface roughness optimization in end milling using the multi objective genetic algorithm approach", *Adv. Mat. Res.,* vol. 576, pp. 103-106, 2012.
[http://dx.doi.org/10.4028/www.scientific.net/AMR.576.103]

[19] S. Jeevanantham, "Effect of machining Parameters on MRR and surface roughness in internal grinding using EN8, EN31 steel", *Int. J. Appl. Eng. Res.,* vol. 12, no. 11, pp. 2963-2968, 2017.

[20] G. Hari Krishna Rao, "Effect of cutting Parameters on the surface roughness of MWCNT reinforced epoxy composite using CNC end-milling process", *Int. J. Sci. Res.,* vol. 2, no. 11, pp. 51-55, 2013.

[21] "Jatin; Pankaj Sharma. Effect of machining Parameters on output characteristics of CNC milling using Taguchi optimization technique. International Journal of Engineering", *Business and Enterprise Applications,* vol. 6, no. 1, pp. 64-67, 2013.

[22] D.S.C. Jayswal, and M. Taufik, "Modeling and analysis of the surface roughness and geometrical error using Taguchi and response surface methodology", *Int. J. Eng. Sci. Technol.,* vol. 3, no. 7, pp. 5809-5815, 2011.

[23] I. Abu Mahfouz, O. El Ariss, A.H.M. Esfakur Rahman, and A. Banerjee, "Surface roughness prediction as a classification problem using support vector machine", *Int. J. Adv. Manuf. Technol.,* vol. 92, no. 1-4, pp. 803-815, 2017.
[http://dx.doi.org/10.1007/s00170-017-0165-9]

[24] S. Kalidass, and P. Palanisamy, "Prediction of surface roughness for AISI 304 steel with solid carbide tools in end milling process using regression and ANN models", *Arab. J. Sci. Eng.,* vol. 39, no. 11, pp. 8065-8075, 2014.
[http://dx.doi.org/10.1007/s13369-014-1346-6]

CHAPTER 6

Analysis on Synthesized Carbon Aerogels Doped with Resorcinol Formaldehyde by XPS and Fesem Techniques

S. Nallusamy[1,*] and **K. Sujatha**[2]

[1] *Department of Adult, Continuing Education and Extension, Jadavpur University, Kolkata-700032, India*

[2] *Department of Chemical Engineering, Dr. M.G.R. Educational and Research Institute, Chennai-600095, Tamilnadu, India*

Abstract: Aerogels are highly porous nanostructures with very low density and have various applications. This research study aims to synthesize carbon aerogels using a sol gel poly-condensation reaction. The primary precursors were resorcinol with formaldehyde in a molar ratio of 1:2 using base catalyst sodium carbonate to produce carbon aerogels. The density obtained for synthesized Resorcinol Formaldehyde (RF) aerogels was 40-50kg/m^3. The carbonization of resorcinol formaldehyde aerogels was carried out using pyrolysis under 800°C with a heating rate of 10°C/min under vacuum conditions after the carbonization porosity of the aerogel was increased. The surface morphology was characterized using Field Emission Scanning Electron Microscopy (FESEM), and elemental analysis was carried out using X-Ray Photoelectron Spectroscopy (XPS). From the final results, it was observed that the carbonization of the resorcinol formaldehyde aerogels by pyrolysis increased the porosity of the carbon aerogels. Thus, it could be predicted that these carbon aerogels can serve as a promising applicant for future water treatment.

Keywords: Aerogel, Density, FESEM, Insert, Molor Ratio, Pyrolysis, Resorcinol Formaldehyde, Super Critical Drying, Water Treatment, XPS.

INTRODUCTION

In a global state, aerogels are nanometer-sized particle arrangements that are well linked to generating foams with low down density strong and elevated surface area. The elements presented in the sol are cross-linked to create a solid three-dimensional set of connections known as gel [1 - 3]. Aerogel was formed when

* **Corresponding author S. Nallusamy:** Department of Adult, Continuing Education and Extension, Jadavpur University, Kolkata-700032, India; E-mail: ksnallu@gmail.com

C. Samson Jerold Samuel, M. Suresh & S. Gnanasekaran (Eds.)

the liquids presented in apertures of the gel were replaced with gas using supercritical drying.

Entity particles in the set of connections are detained collectively with chemical linking so that they become stronger than that produced with the other methods [4 - 6]. Pekala blended the first RF aerogels in 1989 with catalyzed base poly-condensation materials blended on a suitable molar ratio in an aqueous medium using sodium carbonate as a catalyst. Then, the sol is made of wet gel by transferring to a closed container with preset heat for an adequate period. Wet gels are dried with supercritical drying. The organic aerogels of this invention are stable, transparent, low density, and uniform structure pore size with ≤ 0.1micron [7 - 9]. The initial studies describe the different blends with procedures used to develop various gels and explain how these surroundings influence the gel's final composition [10 - 12].

Many articles revealed that procedures to develop different gels and the conditions influence the gel's final composition with the report on the distinctiveness of different properties of organic aerogels and xerogels [13 - 15]. Organic aerogels developed from sol-gel polycondensation of metal alkoxides to examine the microstructural and mechanical properties could be pyrolyzed with an inert ambiance to create carbon aerogel. Results reveal that the carbon and RF aerogels have ultrafine cell/pore size and diameters of 30-200A° [16 - 18]. The organic and carbon porous structure of aerogel was reported, which stated that the aerogel surface area was restricted by the resorcinol mole ratio to the basic catalyst. After preparing the RF aerogel, pyrolysis was performed to prepare the mesoporous carbon aerogel at 1223K. The mesoporous volume was augmented by the pyrolysis of RF aerogel and it was stated that it has a dispersed porous structure [19 - 21]. The structure of organic aerogel with its morphology and scaling were studied. The structure of nano-porous organic aerogel made by the polycondensation of resorcinol with formaldehyde under the based catalyst was determined by small-angle X-ray scattering. The varying concentration of catalyst, monomer concentration, aging, and pyrolysis might cause an effect on the morphology of the aerogel, which was super critically dried. The results reveal that the surface morphology of the rough surface in the length scale was around 10 to 600A° [22 - 24].

This study focuses on the synthesis and characterization of carbon aerogel with varying parameters like catalyst concentration, different molar ratios of catalyst and precursor, aging time, and pyrolysis temperature. Based on this study, an attempt was made to synthesize and characterize RF aerogels derived from organic monomers by FESEM and XPS.

MATERIALS AND METHODS

Carbon aerogels are produced by a base-catalyzed sol-gel process followed by supercritical drying. The chemicals used were Resorcinol (R) (Fisher Scientific, purity 99%), Formaldehyde (F) (Fisher Scientific, purity 41%), and sodium carbonate anhydrous (C) (Fisher Scientific, purity 100%). All solutions were prepared using deionized water (W) and Ethanol was used for the solvent exchange before supercritical drying. The chemical structure of RF aerogel formation is composed of polymer clusters (Fig. **1**).

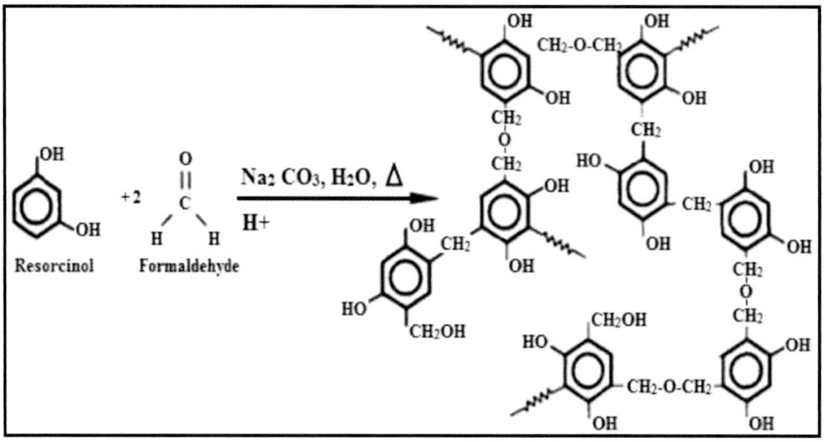

Fig. (1). RF aerogel formation chemical structure [25].

Resorcinol of 1 and 3 dihydroxy benzene and formaldehyde were mixed in a molar ratio of 1:2, respectively. Resorcinol acts as a triserviceable group in 2, 4, and 6th ring positions at the process of polymerization. Formaldehyde is bi-functional and develops a covalent bond among the resorcinol rings. The resulting sol will be composed of polymer clusters, where the pore sizes have been restricted by varying the concentration of base catalyst sodium carbonate. The sol was transferred to glass vials and cured at 80°C. Next, crossed linked RF gels were kept for solvent exchange in ethanol and subsequently dried by supercritical drying optimization of the synthesis was conducted by varying the concentration of catalyst and the period for gelation. The optimized synthesis parameters were up-scaled to 8 times the base volume.

RESEARCH METHODOLOGY

The preparation methodology of RF aerogel is shown (Fig. **2**) as a flow chart. The carbon aerogels were developed from the pyrolysis of RF aerogels in a tubular furnace. The furnace was loaded with the aerogel sample, subsequently evacuated, and filled with nitrogen. Nitrogen was poured into the furnace, heated at 10°C/min to 800°C/min, and detained at the same heat for an hour. Then, the furnace could be cooled at room temperature. The wet gels were treated with fluoride water of 5ppm prepared from NaF standard stock of 1000ppm and where the fluoride concentration was analyzed by water test kit procured from NICE chemicals, Chennai.

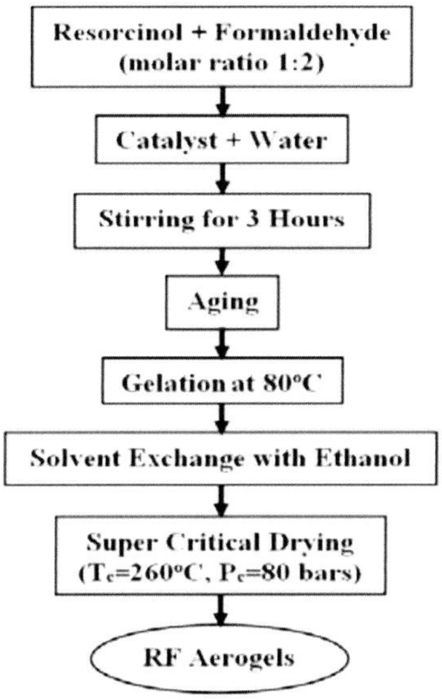

Fig. (2). RF aerogel preparation methodology.

Aerogels densities were determined by calculating the weight and volume of the samples. FESEM was used to scan the shell morphologies of resorcinol formaldehyde aerogel. The RF samples were coated with gold by sputtering before imaging. The ZEISS Gemini-500 scanning electron microscope with a beam voltage was used to capture secondary electron images of aerogels cross-sections. The elemental composition of RF aerogel was determined with an elemental analyzer and the type of surface functional groups were analyzed by XPS performed on an Omicron nanotechnology. Dried wet gels treated with fluoride water were analyzed for the surface adsorption of fluoride by RF gel.

RESULT AND DISCUSSION

Initially, the scale-up of RF gels with the same base parameters got precipitated rather than forming a gel. So this was resolved by increasing the interaction between the catalyst and precursor by increasing the aging time of the sol. The supercritical drying resorcinol formaldehyde aerogels were produced with dark red color and found to be soft, brittle, and elastic. The carbonization of the RF aerogel using pyrolysis was observed to be black. This infers the complete transformation of the RF network into a carbon network. It was also found that there is a decrease of 40% in the weight of the carbon aerogel with its respective RF aerogel, which further confirmed the removal of other functional groups in the RF network.

Determination of Density

The synthesized RF aerogel was very light weighted and had a density around 50-60Kg/m^3. The density of carbonized was decreased by approximately half of the respective RF aerogels and found to be 25-30Kg/m^3 after calculation.

FESEM Study

The comparative study on the surface morphology of small-scale and upscale RF aerogels formed a similar continuous structural pattern. There was not much difference in pore size of 40-50A° and with high porosity nature in both small scale (Fig. **3**) and upscale (Fig. **4**) RF aerogels. There was a significant increase in the porosity in carbon aerogel compared to RF Aerogels (Fig. **5**). This can confirm the removal of the other groups except carbon (Fig. **6**) from the RF network.

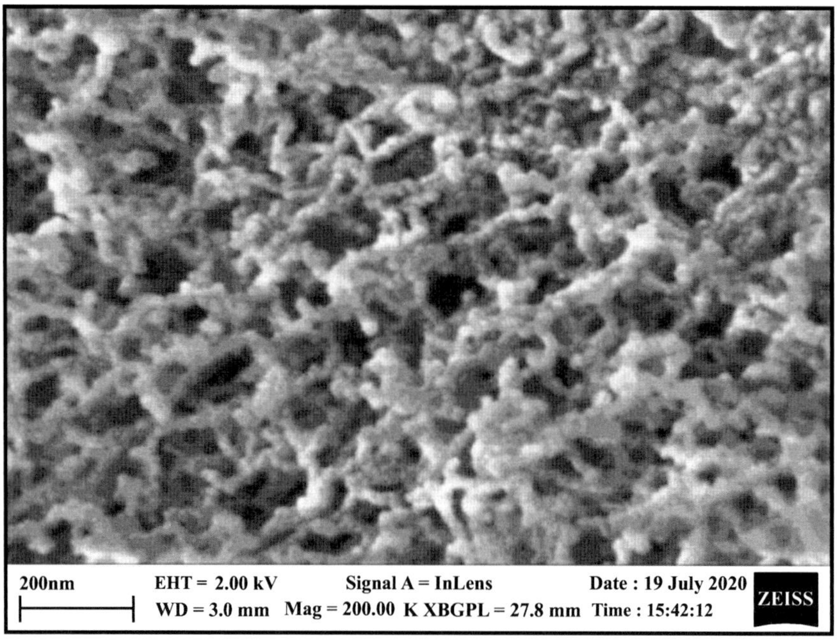

Fig. (3). Surface morphology of small scale RF aerogels.

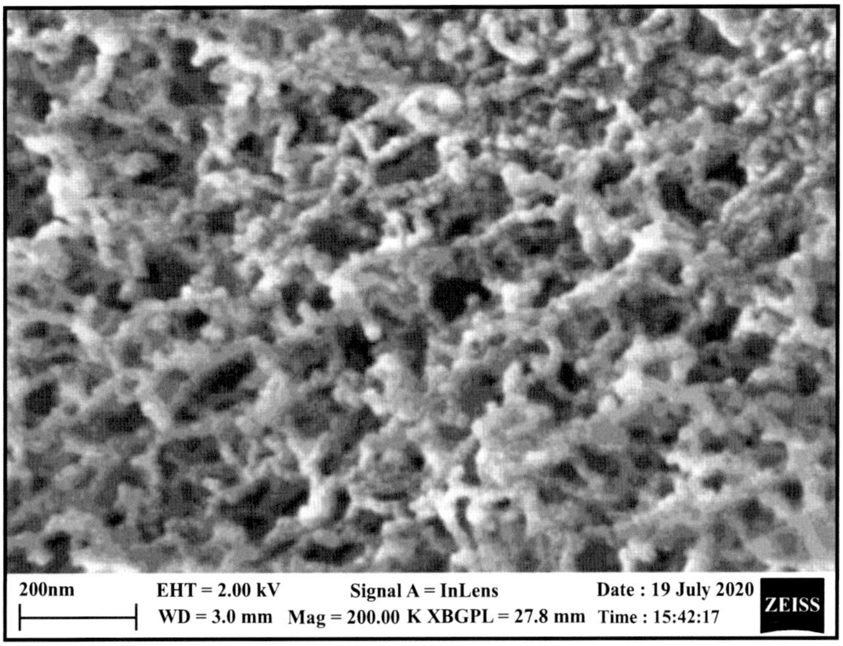

Fig. (4). Surface morphology of upscale RF aerogels.

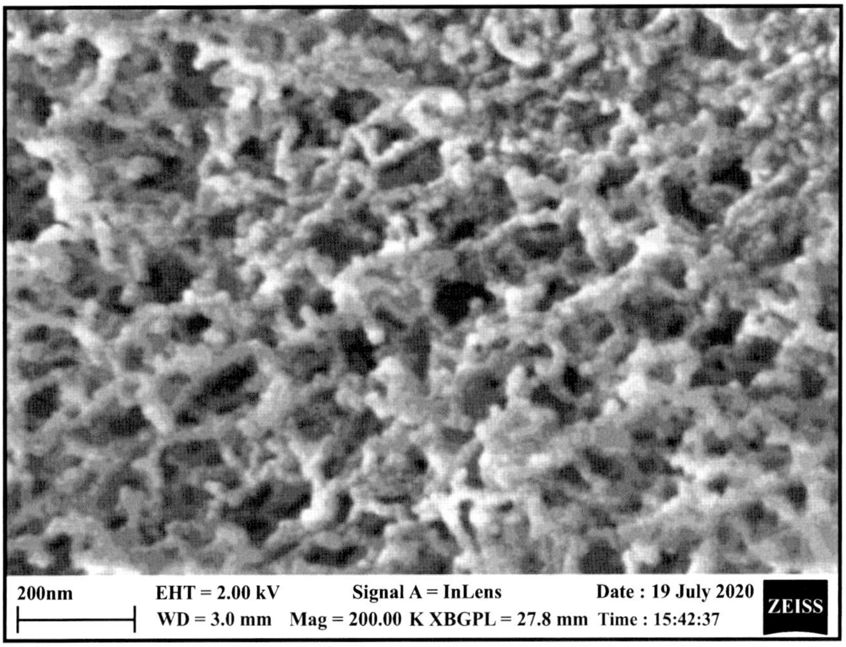

Fig. (5). Comparative study SEM image of RF aerogel.

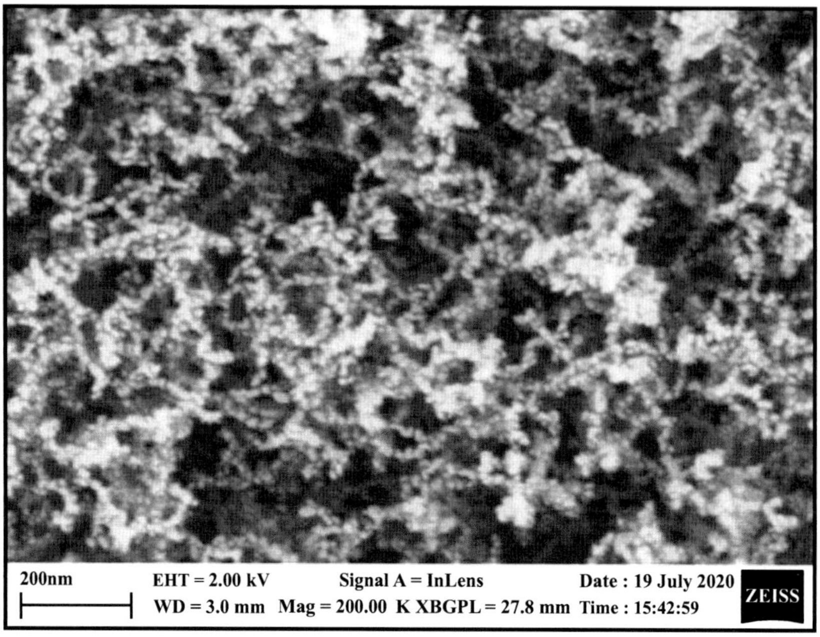

Fig. (6). Comparative study SEM image of carbon aerogel.

XPS Study

The plots for the total survey of RF aerogel with insert a) O 1s Spectra and b) C 1s Spectra (Fig. **7**) and survey of RF aerogel with insert a) Na 1s Spectra and b) F 1s Spectra (Fig. **8**) are shown respectively. The total survey of element analysis found the presence of carbon and oxygen peaks in 532kV and 234kV, respectively. Moreover, there were 4 distinct peaks of carbon in the same binding energy range 283kV to 290kV could be the different bondings of carbon with oxygen and neighboring carbon atoms, which have to be studied further. Similarly, oxygen also showed three distinct peaks that would support the hypothesis of oxygen bonding to hydrogen as well as carbon. The XPS spectra of fluoride-treated RF gels showed a distinct binding energy peak of fluoride inside. This confirms the surface adsorption capacity of the RF gel on fluoride.

Text are not readable please provide clear text

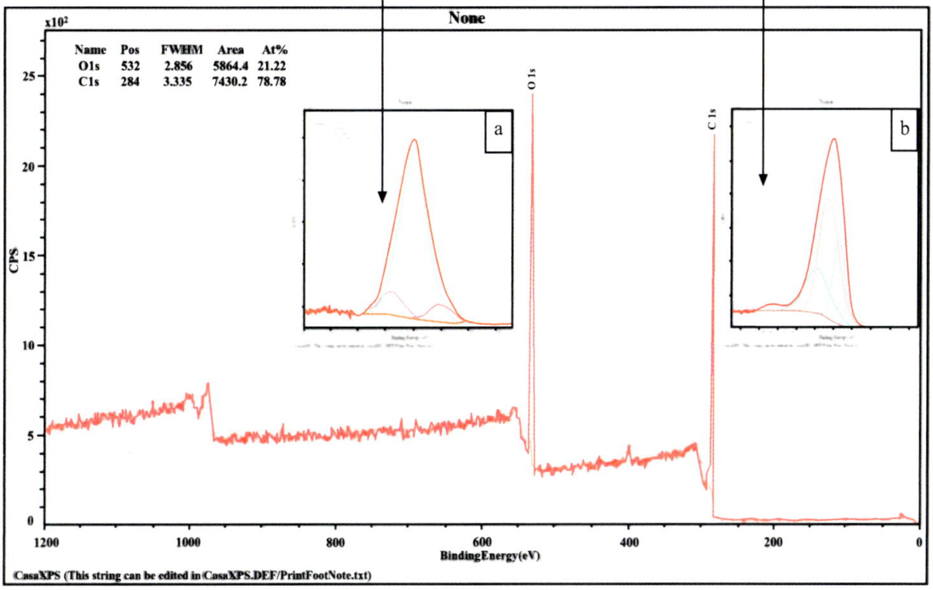

Fig. (7). XPS results of RF aerogel (insert: a) O1 s spectra and b) C 1s spectra).

Text are not readable please provide clear text

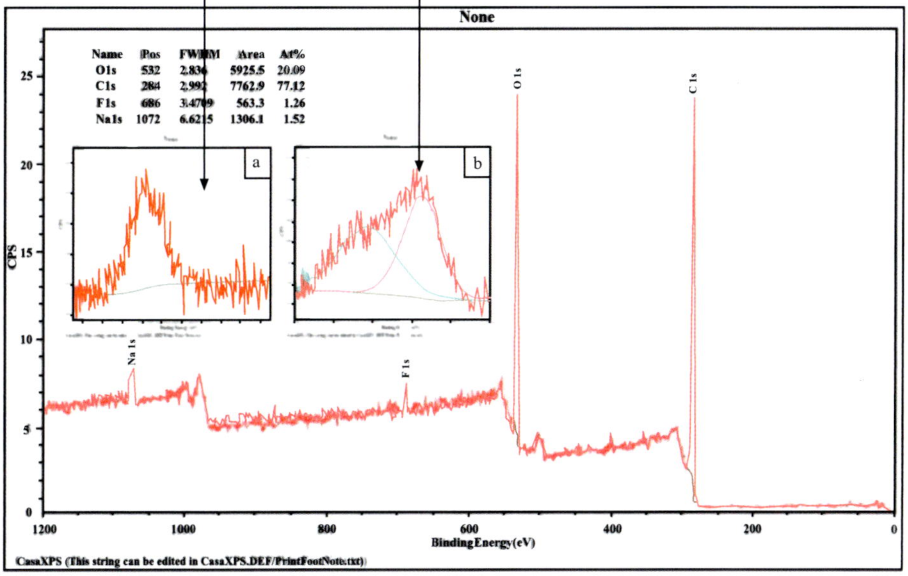

Fig. (8). XPS results of RF aerogel (insert: a) Na 1s spectra and b) F 1s spectra).

CONCLUSION

In this experimental study, carbon aerogels were synthesized successfully from sol-gel polymerization of resorcinol with formaldehyde. The following conclusions were derived from the results obtained:

- RF aerogels were solid matrices composed of interconnected particles with a diameter of 40-50A°.
- The scale-up of RF was studied and the significance of aging was found in the gelation.
- The carbonization of the RF gels by pyrolysis increased the porosity of the carbon aerogels.
- Hence, it could be envisaged that these carbon aerogels can serve as a promising candidate for future water treatment.

In future studies, more attention is needed for further research to get a new path with the novel idea to accomplish the significant applications in this research area.

CONSENT FOR PUBLICATION

Not applicable.

CONFLICT OF INTEREST

The author declares no conflict of interest, financial or otherwise.

ACKNOWLEDGEMENTS

Declared none.

REFERENCES

[1] F. Despite, S. Calas Etienne, and P. Etienne, "Slow crack growth in silica aerogels: A review", *J. Sol-Gel Sci. Technol.*, vol. 90, no. 1, pp. 20-27, 2019. [http://dx.doi.org/10.1007/s10971-018-4857-x]

[2] Zhencheng Yu, "Cellulose nanofibril/carbon nanomaterial hybrid aerogels for adsorption removal of cationic and anionic organic dyes", *Nanomaterials,* vol. 10, no. 1, pp. 1-20, 2020.

[3] H. Zhang, J. Feng, L. Li, Y. Jiang, and J. Feng, "Controlling the microstructure of resorcinol–furfural aerogels and derived carbon aerogels *via* the salt templating approach"., *RSC Advances,* vol. 9, no. 11, pp. 5967-5977, 2019. [http://dx.doi.org/10.1039/C9RA00238C]

[4] M. Schwan, and L. Ratke, "Lorenz Ratke. Flexibilization of resorcinol formaldehyde aerogels", *J. Mater. Chem. A Mater. Energy Sustain.,* vol. 1, no. 43, pp. 13462-13468, 2013. [http://dx.doi.org/10.1039/c3ta13172f]

[5] C. Bhan, J. Singh, and Y.C. Sharma, "Development of adsorbent from Mentha plant ash and its application in fluoride adsorption from aqueous solution: a mechanism, isotherm, thermodynamic, and

kinetics studies", *Int. J. Phytoremediation,* vol. 23, no. 11, pp. 1113-1123, 2021. [http://dx.doi.org/10.1080/15226514.2021.1880365] [PMID: 33561355]

[6] P. Kumar, S. Kumar, and S. Suganya, "Treatment of fluoride contaminated water- A review", *Environmental Chemistry Letters,* vol. 17, no. 4, pp. 1707-1726, 2019. [http://dx.doi.org/10.1007/s10311-019-00906-9]

[7] H. Maleki, "Recent advances in aerogels for environmental remediation applications", *Chem. Eng. J.,* vol. 300, pp. 98-118, 2016. [http://dx.doi.org/10.1016/j.cej.2016.04.098]

[8] M. Salihovic, N. Hüsing, J. Bernardi, V. Presser, and M.S. Elsaesser, "Carbon aerogels with improved flexibility by sphere templating", *RSC Advances,* vol. 8, no. 48, pp. 27326-27331, 2018. [http://dx.doi.org/10.1039/C8RA04848G]

[9] R. Araga, Kali Suresh, and S. Sharma Chandra, "Coconut shell derived carbon/carbon nanotube composite for fluoride adsorption from aqueous solution", *CLEAN–Soil, Air, Water,* vol. 47, no. 5, p. 1800286, 2019. [http://dx.doi.org/10.1002/clen.201800286]

[10] A. Lamy-Mendes, R.F. Silva, and L. Durães, "Rui F. Silva; Luisa Duraes. Advances in carbon nanostructure–silica aerogel composites: a review", *J. Mater. Chem. A Mater. Energy Sustain.,* vol. 6, no. 4, pp. 1340-1369, 2018. [http://dx.doi.org/10.1039/C7TA08959G]

[11] M. Guild, "On the use of aerogel as a soft acoustic metamaterial for airborne sound", *Physical Review Applied,* vol. 5, no. 3, pp. 1-12, 2016.

[12] M. Vithanage, and P. Bhattacharya, "Fluoride in the environment: sources, distribution and defuoridation", *Environ. Chem. Lett.,* vol. 13, no. 2, pp. 131-147, 2015. [http://dx.doi.org/10.1007/s10311-015-0496-4]

[13] T. Akai, "Removal of fluoride from drinking water by sorption using diatomite modified with aluminum hydroxide", *Journal of Analytical Methods in Chemistry,* pp. 1-11, 2019.

[14] N. Alena, "Electrochemical properties of carbon aerogel electrodes: dependence on synthesis temperature", *Molecules,* vol. 24, pp. 1-13, 2019.

[15] M. Thakkar, Z. Wu, L. Wei, and S. Mitra, "Water defluoridation using a nanostructured diatom-ZrO2 composite synthesized from algal biomass", *J. Colloid Interface Sci.,* vol. 450, pp. 239-245, 2015. [http://dx.doi.org/10.1016/j.jcis.2015.03.017] [PMID: 25823727]

[16] Woignier Thierry, "Techniques for characterizing the mechanical properties of aerogels", *Journal of Sol gel Science and Technology,* vol. 93, pp. 6-27, 2020.

[17] M.L. Oyen, and R.F. Cook, "A practical guide for analysis of nanoindentation data", *J. Mech. Behav. Biomed. Mater.,* vol. 2, no. 4, pp. 396-407, 2009. [http://dx.doi.org/10.1016/j.jmbbm.2008.10.002] [PMID: 19627846]

[18] S. Chandrasekaran, B. Yao, T. Liu, W. Xiao, Y. Song, F. Qian, C. Zhu, E.B. Duoss, C.M. Spadaccini, Y. Li, and M.A. Worsley, "Direct ink writing of organic and carbon aerogels", *Materials Horizons,* vol. 5, no. 6, pp. 1166-1175, 2018. [http://dx.doi.org/10.1039/C8MH00603B]

[19] J.P. Vareda, and L. Durães, "Efficient adsorption of multiple heavy metals with tailored silica aerogel-like materials", *Environ. Technol.,* vol. 40, no. 4, pp. 529-541, 2019. [http://dx.doi.org/10.1080/09593330.2017.1397766] [PMID: 29098957]

[20] A. Faivre, L. Duffours, P. Colombel, and F. Despetis, "Mechanical behaviour of aerogels and composite aerogels submitted to specific penetration tests. Journal of Sol gel", *Sci. Tech. (Paris),* vol. 90, pp. 67-75, 2019.

[21] X. Yue, J. Chen, H. Li, Z. Xiao, X. Yu, and J. Xiang, "One pot rapid synthesis of ultrahigh strength

hydrophobic bulk silica aerogels", *Mater. Chem. Front.,* vol. 4, no. 8, pp. 2418-2427, 2020. [http://dx.doi.org/10.1039/D0QM00234H]

[22] M. Suneetha, B. Syama Sundar, and K. Ravindhranath, "Studies on defuoridation techniques: A critical review", *Int. J. Chemtech Res.,* vol. 8, pp. 295-309, 2015.

[23] N. Rustagi, A.S. Rathore, J.K. Meena, A. Chugh, and R. Pal, "Neglected health literacy undermining fluorosis control efforts: A pilot study among schoolchildren in an endemic village of rural Rajasthan, India", *J. Family Med. Prim. Care,* vol. 6, no. 3, pp. 533-537, 2017. [http://dx.doi.org/10.4103/2249-4863.222017] [PMID: 29417003]

[24] N.K. Mondal, R. Bhaumik, and J.K. Datta, "Removal of fluoride by aluminum impregnated coconut fiber from synthetic fluoride solution and natural water", *Alex. Eng. J.,* vol. 54, no. 4, pp. 1273-1284, 2015. [http://dx.doi.org/10.1016/j.aej.2015.08.006]

[25] A. Allahbakhsh, and A.R. Bahramian, "Self-assembled and pyrolyzed carbon aerogels: an overview of their preparation mechanisms, properties and applications", *Nanoscale,* vol. 7, no. 34, pp. 14139-14158, 2015. [http://dx.doi.org/10.1039/C5NR03855C] [PMID: 26245296]

CHAPTER 7

Tribological Behaviour of Hybrid Natural Fiber Polymer Composites with Novel Fillers

P. Pitchipoo[1,*], A. Muthiah[1], A. Manikandan[2] and S. Rajakarunakaran[3]

[1] *Mechanical Engineering, P.S.R. Engineering College, Sivakasi, India*

[2] *Mechanical Engineering, Lord Jagannath College of Engineering and Technology, Nagercoil, India*

[3] *Mechanical Engineering Ramco Institute of Technology, Rajapalayam, India*

Abstract: This study has been carried out to demonstrate the tribological behavior of natural fiber-based polyester hybrid composite with novel filler materials. The reinforcement fibers considered in this study are Jute and Coir fiber. The filler materials selected in this work are copper slag (final waste from the copper extraction) and Nanoclay (Montmorillonite K10). The percentage of both reinforcement fibers and fillers has been changed, and the analyses have been carried out. Experiments were conducted on the specimen based on Taguchi's design of experimental technique. L9 Orthogonal array is selected for analysis, and Grey Relational Analysis (G.R.A.) is introduced to determine the better combination of reinforcement and filler to provide better wear resistance and coefficient of friction. Also, a statistical analysis is performed to find the influence of input parameters.

Keywords: ANOVA, Coefficient of Friction, Coir Fiber, Composite Material, Compression Moulding, Copper slag, Grey Relational Analysis, Hybrid Fiber, Hybrid Filler, Jute Fiber, Multi-Objective Optimization, Nanoclay, Natural Fiber, Orthogonal Array, Parameter Optimization, Pin on Disc Wear Test, Polymer Composites, Taguchi Technique, Tribological behavior, Wear.

INTRODUCTION

There is a greater demand for materials with unusual combinations of properties in the present situation. It is required that the material must have low density, high strength, abrasion-resistant, impact resistance, and increased hardness [1]. It is difficult to obtain all the properties in the same material. So the research has been carried out, and the engineers have tried to develop a material taking all the properties into consideration. This hunt for various increased properties in a

* **Corresponding author P. Pitchipoo:** Mechanical Engineering, P.S.R. Engineering College, Sivakasi, India; E-mail: drpitchipoo@gmail.com

C. Samson Jerold Samuel, M. Suresh & S. Gnanasekaran (Eds.)

material led to the development of composite materials. Composite materials are materials that have two or more phases that are used to enhance the properties. The composite consists of a load-carrying material with matrix and filler elements. The most commonly used matrix materials are varied based on the type of composites; a few of them are metal, ceramics, and polymers. Our research has employed polymer as the matrix substance for their low specific gravity, low density, and chemical stability. The application can be mainly concentrated in automotive industries for various requisites. There are various types of composites based on the variations in their matrix materials and reinforcement materials, such as fiber-reinforced composites, dispersion strengthened composites, and particle reinforced composites.

The natural fibers available on the earth are bio-degradable materials. They hold numerous properties and enormous applications. One key application is the use of natural fiber as reinforcement in fiber composites. Natural fibers can be effectively used as synthetic fiber in fiber-reinforced polymers. Natural fiber reinforcement can increase the mechanical properties of polymers, which is more economical and eco-friendly. Two major drawbacks of matrix and synthetic reinforcement materials can be eliminated by using natural fibers as reinforcement [1]. John and Thomas [2] clarified that natural fibers could be developed by plants, animals, and geological processes. Vegetable fibers (cotton, jute, coir, hemp, flax, ramie, sisal, bagasse, banana, *etc.*) commonly have cellulose and lignin. Plant fibers are engaged in producing paper, textile, and dietary fiber is a vital element of human nutrition. Coir fibers are seen between the harder inner shell and the outer skin of a coconut. The individual fiber cells are slim and hollow, with cellulosic walls [3]. Jute is one of the most important plant fibers (Lignocellulosic fiber) available worldwide. It is drawing attention as potential reinforcement of polymer composites because of its innate properties such as low density and low elongation at break and high tensile modulus. Moreover, its specific stiffness and strength are similar to synthetic fibers. It has been found that natural fiber composites possess excellent electrical resistance, good thermal and acoustic insulating properties, and high resistance to fracture [4].

Solid wastes from the industries are generated as waste residue, mining waste, ore dressing tailings, smelting, and chemical processes. At present, these wastes are managed by storage methods and discharge methods, comprehensive utilization, and disposal [5]. One such waste is copper slag which is an industrial waste generated during the production of copper. Nanoclay, also known as Montmorillonite, is a soft silicate cluster of minerals formed from the sediment in water solution as tiny crystals. Nanoclay increases materials' strength, fatigue resistance, stiffness, and thermal stability [6]. The more aspect ratio of nanoclay results in a superior interfacial area, which significantly increases its reinforcement properties.

This study attempts to develop a hybrid natural fiber polymer matrix composite material with novel filler to get better tribological properties.

LITERATURE REVIEW

Mukherjee *et al.* [7], Satyanarayana *et al.* [8], and Lilholt *et al.* [9] originated research on the structural property studies of coir fibers. Fibers obtained from various parts of the coconut tree were tested for their size, density, resistance to electricity, ultimate tensile strength, tensile modulus, and percentage elongation. The stress-strain graph of the fracture mode, microfibrillar angle as well as cellulose and lignin substances of these fibers were observed in their research. The investigation of Rout *et al.* [10] proved that the untreated coir fibers have weak interfacial bonding with the polyester matrix and the mechanical properties of the composites are poor. Santafe Junior *et al.* [11] investigated that the tensile properties do not improve when the addition of coir fiber increases more than 40% of volume fraction. Ramanaiah *et al.* [12] investigated that when the volume fraction of the fiber is increased, the tensile strength is increased in the fiber-reinforced polyester composites. Monteiro *et al.* [13] evaluated structural characteristics and mechanical properties of coir fiber-polyester composites. Variation in flexural strength with the mass fraction of coir fiber and molding pressure was also analyzed. The outcomes attained for flexural strength permitted the comparison of the technical concert of the composites with other conventional materials. Jayabal and Natarajan [14] investigated natural fiber-reinforced composites, and they highlighted that the coir fiber composites act as a substitute for synthetic fiber composites. The lesser cost and high-modulus weight ratio also suggest the consumption of coir fiber as reinforcement material in composite materials.

Messadi *et al.* [15] made a comparison of flax fiber and glass fiber. The flax fiber has more strength than glass fiber due to the local misalignment of cellulose microfibrils inside the fiber. Flax fiber also has good fracture strength than glass fiber due to its concentric rings of microporous structure. The mechanical properties of thermally treated flax fiber composites possess better mechanical properties and improve the bonding between the interfaces [16]. The stiffness of the flax fiber composites varies due to its unstable circumstance of stress-strain during starting of the loading. The physicochemical properties of the flax fiber are important factors in affecting the mechanical properties [17]. Braga and Magalhaes [18] studied the moisture absorption properties of jute and glass reinforced epoxy composites. Increasing the jute fiber content in the composites increased the moisture absorption property more than that of glass fiber composites.

Hybridization in polymer composites evolved almost two decades earlier. Hybridization in polymer brings increased properties. In hybridization, polymer composites were fabricated by reinforcing different fiber materials together in a single matrix. This concept was first attempted by developing the glass and carbon fiber composite. The main advantage of natural fiber hybridization is the possibility of getting strength comparable to synthetic fibers. Numerous research has been conducted on hybrid composites fabrication and characterization in recent years. Natural fibers were used with synthetic fibers to tailor the properties., *e.g.*, carbon fiber composite shows remarkable tensile characteristics but has poor toughness. This was eliminated by the hybridization of carbon fiber with natural fiber [19]. Nisini *et al.* [20] fabricated the epoxy hybrid composite with carbon, basalt, and flax fibers. The addition of flax fiber between the basalt, and carbon fiber increased the bending characteristics of the composite with high flexural strength. Fragassa *et al.* [21] studied the tensile and impact behavior of the basalt flax hybrid composite and found that the flax fiber reinforcement increased the basalt fiber composites tensile and impact strengths. Three different P.L.A. hybrid composites were fabricated using bamboo, kenaf, and coir fibers. The composite with all three fibers showed higher tensile strength than the bamboo-coir and kenaf-coir composite, which was 20% and 78% higher, respectively. The flexural modulus of the kenaf-coir composite was approximately 70% higher than the other two composites [22]. Banana sisal epoxy composites were fabricated by setting constant (40%) fiber weight percentage, and the tensile properties were investigated [23]. The maximum tensile strength was noted for the composite with 20% sisal fiber and 20% banana fiber. Jawaid *et al.* [24] reported that the hybridization of jute fiber and oil palm fiber increased the tensile and dynamic mechanical properties of the composites. Venkata Subba Reddy *et al.* [25] reported that hybridization of bamboo with glass fiber in polyester improved the chemical resistance of the composite. Hybridization could also lead to negative effects on the mechanical properties. Sreekala *et al.* [26] and Hariharan and Khalil [27] reported the adverse impact of hybridization on tensile strength and young's modulus of the composites.

The tribological properties of the composites were significantly influenced by many factors such as type of fiber, fiber orientation, fiber loading, fiber length, and type of matrix [28, 29]. Gupta *et al.* [30] investigated the frictional coefficient hybrid sisal/jute fiber reinforced epoxy composites with different sliding speeds such as 1-3 m/s with the different loading 10-30 N. They found that reduction in specific wear rate and coefficient of friction due to the incorporation of fibers into the epoxy matrix was present in all conditions. They also found that alkali treatment of fibers into hybrid composite further reduced the value of specific wear-rate and co-efficient of friction. Omrani *et al.* [31] analyzed the wear behaviour of polymer matrix composites reinforced with natural fibers. The outcomes proved that the amalgamation of natural fiber such as jute, kenaf, sisal, coir, rice husk, date and oil palm, sugarcane, and bio-waste products boosted the wear behaviour significantly. After incorporating functional fillers and reinforcing, polymer composites had excellent friction and wear performance. This benefit made them flexible in numerous engineering applications [32, 33]. Cheng and Kortschot [35] reviewed the friction co-efficient of Non woven and woven carbon fiber/epoxy composites in their studies. They found the sliding direction causes in the co-efficient of friction. The low value of frictional co-efficient was identified in the sliding direction parallel to the fiber direction.

Chand and Dwivedi [34] investigated the wear behaviour of jute fiber reinforced polypropylene composites. The effects of coupling agent (maleic anhydride-grafted polypropylene) addition, sliding distance, and load on the abrasive wear performance of the composites were found using an abrasion tester. It was revealed that the addition of a coupling agent during the melt mixing stage imparted better wear resistance to the composites. S.E.M. was used to analyze the abrasive wear performance of the composites. El-Tayeb [35] determined the friction coefficients and wear rates of sugarcane fiber reinforced polyester and glass fiber reinforced polyester composites under dry sliding contact conditions. It was found that polyester composite reinforced with short sugarcane fiber exhibited better wear resistance and friction coefficient equivalent to that of polyester composite reinforced with glass fiber. This performance was attributed to the non-corrosive nature of the sugarcane fiber, which facilitated better fiber-matrix bonding. The wear rate decreased linearly with fiber length and then started increasing at higher values (>= 10 mm). Deepak *et al.* [36] analyzed the mechanical and wear properties of jute/banana fiber reinforced molybdenum disulfide modified epoxy composites. Molybdenum disulfide was added to the epoxy resin at various weight fractions to increase the wear properties of the composites. It was found that abrasion loss of the composites decreased with an increase in the weight fraction of molybdenum disulfide. At the same time, enhancement in wear resistance was noted.

Hasan *et al.* [6] used long fiber Bangla Tossa special jute as reinforcement material along with 1, 3, and 5% nanoclay for preparing polyester composites. Temperature-induced weight loss due to thermal decomposition was also measured, and char residue was calculated at 1000 °C. The 5% nanoclay added samples showed better thermal stability. On the other hand, 1% nanoclay added composites showed better viscoelastic properties (E‘ and E‖) and better stability. In the present scenario, the economic growth of a nation highly depends on the degree of industrialization. Although rapid industrialization signifies development, the wastes from the industries adversely affect the environment and organic life due to their toxic behavior. The growing awareness about global environmental issues demands the potential management of industrial wastes. Wastes produced worldwide are classified into three major divisions: industrial wastes, commercial wastes, and construction wastes. As reported worldwide, nearly 7-10 billion tons of urban wastes are generated every year, in which industrial wastes occupy 21% of the total [37]. Rajini *et al.* [38] investigated the effect of nanoclay addition on the free vibration and mechanical properties of unsaturated polyester hybrid composites reinforced with naturally woven coconut sheath/E-glass/nanoclay. It was observed that the addition of nanoclay considerably improved the mechanical and damping behaviour of the composite. This enhancement was attributed to the improved stiffness of the matrix-induced by better dispersion of nanoclay. Saba *et al.* [39] studied the effects of adding various nanofillers (oil palm nanofiller, MMT, and organically modified MMT) on the physicomechanical properties of kenaf fibers reinforced epoxy hybrid composites. Upon nanofiller loading, the density of the hybrid composites increased, which enhanced the hardness property. The uniform distribution of nanofiller in the matrix was confirmed by XRD analysis. All the nanofiller-loaded composites exhibited lower values of coefficient of thermal expansion. The improvement in the properties was attributed to the better fiber-matrix interfacial bonding due to the presence of nanoclay facilities. This urged researchers to develop new, improved techniques for the proper storage and disposal of these wastes. At present, these wastes are stored in large lagoon-type impoundments. Alkalinity and sodicity are the two major characteristics of the red mud responsible for its toxic behavior [40]. When dumped into the impoundments, this toxic waste affects the surrounding soil, nature, and groundwater.

It becomes more necessary to eliminate these wastes by developing new materials. Also, proper utilization of these wastes can provide a cleaner production and sustainable environment. Polymer composites are promising materials that replace most conventional materials in various applications. In recent years, polymer composites have been fabricated using filler reinforcement. Various industrial wastes have been utilized as reinforcements in polymers by many researchers. Industrial waste like rice husk, fly ash, sawdust, and cottage industry wastes are successfully used as potential reinforcing materials in the polymer matrix and studied for their mechanical properties and other characteristics [41]. The addition of these wastes offers an inherent advantage in industrial waste management, cost-effective material fabrication, and realizing acceptable mechanical properties in the developed materials. Red mud has been used in various material systems as filler reinforcement and studied for its properties. Red mud reinforced metal matrix composites were fabricated and investigated for dry sliding wear properties.

The results concluded that wear resistance is improved with the addition of red mud reinforcement. Liu and Poon [42] utilized red mud in self-compacting concretes and studied the compressive and tensile splitting properties. Ye *et al.* [43] manufactured a geo-polymer using red mud and noted the enhancement in compression and long-term strength. Hybrid composite was fabricated by Arumugaprabu *et al.* [44] using red mud, and the mechanical, damping, and chemical resistance properties were investigated. The work done by Biswas and Satapathy [45] reported the tribological behavior of industrial waste-filled epoxy composites. Red mud waste was used as a filler in the glass fiber matrix. The work also recommends the possible use of developed composite material for industrial fans, housing, fiber boats, sports kits, structures, *etc.* G.R.A. was used to solve multi-objective problems like parameter optimization in GMAW [46], mechanical properties of composite materials [47], *etc.* In this work, the tribological properties of natural fiber composite material with novel fillers are optimized.

MATERIALS AND METHODS

Materials Used

In this study, polyester resin is used as a matrix material, and coir fiber and jute fiber are selected for reinforcement. Nanoclay and copper slag are chosen as filler materials. The basic properties of the chosen materials are presented in Table **1**. Jute Fiber is a naturally available fiber that comprises bundled ultimate cells, each containing spirally oriented micro- fibrils bound together. Coir fiber is a natural fiber obtained from coconut fiber that is pulled out from the shell of the coconut and has application in the products like floor mats, doormats, brushes, mattresses, *etc.*

Table 1. Basic Properties.

Properties	Polyester Resin	Jute Fiber	Coir Fiber	Copper slag	Nanoclay
Density (g/cm3)	1.1 to 1.43	1.3-1.45	1.15	2.31	1.98
Tensile Strength (MPa)	34.5 – 103.5	393-773	131-175	-	-
% Elongation	1 - 5	1.16-1.5	15-40	-	-

Copper slag is a derivate of copper extraction from its ore by smelting. During smelting, impurities develop slag, which drifts on the molten metal. The ingredients of copper slag are shown in Table **2**.

Table 2. Copper slag Composition.

Ingredients	Percentage
SiO_2	33-35%
Fe_2O_3	40-44%
Al_2O_3	4-6%
CaO	0.8-1.5%
MgO	1-2%

Montmorillonite is a very soft phyllosilicate group of minerals that typically form as microscopic crystals known as Nanoclay. It is one of the most extensively explored catalytic minerals in heterogeneous catalysis due to its low cost and eco-friendliness. It is a hydrophilic property and has a surface area of 250 sq. m/g. It can be reinforced to enhance mechanical strength and physical properties by forming mixed matrix membranes embedded with an inorganic phase. MMT can enhance the selectivity and permeability of the membranes and make the matrix set with reinforcement.

Methodology Used

To run the experiments, Taguchi's design of experiments concept is adapted in this work. This technique is used to reduce the deviation in a process through the robust design of experiments. The prime objective of the technique is to make a product with optimal quality and cost. The Taguchi technique was established by Ganache Taguchi. The experimental design suggested by Taguchi uses orthogonal arrays to utilize the factors affecting the process and the lower and upper limits at which they can be varied. In this study, three various parameters with three levels are considered controlling factors. They are the combination of fiber, percentage of copper slag, and percentage of nanoclay (Table **3**).

Table 3. Parameters and Levels.

	Level 1	Level 2	Level 3
Fiber (A)	Jute	Coir	Jute + Coir
Copper Slag (gm) (B)	5	2.5	0
Nano – Clay (gm) (C)	0	2.5	5

Actually, 27 combinations can be obtained using these three factors with three levels combinations. This incurs more cost and time consuming also. To rectify these issues, Taguchi suggested only 9 experiments based on the L9 orthogonal array. The L9 orthogonal array and the respective combinations of the parameters are presented in Table **4**. The specimens are fabricated using these combinations, and the experiments are performed.

Table 4. Standard L9 Table.

Experiment No.	A	B	C	A	B	C
1	1	1	1	Jute	5	0
2	1	2	2	Jute	2.5	2.5
3	1	3	3	Jute	0	5
4	2	1	2	Coir	5	2.5
5	2	2	3	Coir	2.5	5
6	2	3	1	Coir	0	0
7	3	1	3	Jute + Coir	5	5
8	3	2	1	Jute + Coir	2.5	0
9	3	3	2	Jute + Coir	0	2.5

Grey Relational Analysis (G.R.A.)

Taguchi method can solve a single-objective optimization problem. But the problem considered is a multi-objective problem. The objectives expected are minimum wear loss and maximum coefficient of friction. So G.R.A. is used in this work to attain the objectives [48]. Initially, S/N ratio is determined to eliminate the influence of external factors during the experimentation. S/N ratios are calculated based on the smaller, the better [Eqn. (1)] and the larger, the better concepts [Eqn. (2)], depending on the quality parameters.

$$\frac{S}{N} = -10\log\left(\sum_{i=1}^{n} y^2\right) \quad \textbf{(1)}$$

$$\frac{S}{N} = -10\log\left(\sum_{i=1}^{n} \frac{1}{y_i^2}\right) \quad \textbf{(2)}$$

where i = experiment number, n = number of experiments, and yi= value of the parameter for the ith experiment.

The steps below were followed while applying G.R.A.

Step 1: Data Pre-processing

Step 2: Calculation of the grey relational coefficient

Step 3: Calculation of the grey relational grade

Step 5: Select the optimal levels of process parameters

Data Pre-processing

In the study, a linear data preprocessing method for the output response is the higher-the-better and is expressed as

$$x_i\ (k) = \frac{y_i\ (k) - \min y_i\ (k)}{\max y_i\ (k) - \min y_i\ (k)} \quad \textbf{(3)}$$

If the expected output is minimum, that is preprocessed using the lower-the-better [Eqn. (4)] concept.

$$x_i\ (k) = \frac{\max y_i\ (k) - y_i\ (k)}{\max y_i\ (k) - \min y_i\ (k)} \quad \textbf{(4)}$$

Where x_ik is the value after the grey relational generation, min y_ik is the smallest value of y_ik for the kth response, and max y_ik is the largest value of y_ik for the kth response.

Calculation of Grey Relational Co-efficient

The definition of the grey relational grade in the grey relational analysis is to show the relational degree between the nine sequences [$x_0(k)$ and $x_i(k)$, = 1, 2, . . ., 9]. The grey relational coefficient $\xi i(k)$ can be calculated by using [Eqn. (5)].

$$\xi_i(k) = \frac{\Delta_{min} + \zeta\Delta_{max}}{\Delta_{0i}(k) + \zeta\Delta_{max}} \quad \textbf{(5)}$$

Where $\Delta_{0i} = ||x_0\ (k) - x_i\ (k)||$ is the difference of the absolute value between x_0 (k) and x_i (k); ζ is distinguishing coefficient between 0 ~ 1 Δ; $\forall j^{min}||x_0(k) - x_j(k)||$ is the smallest value of Δ_{0i} ; $\Delta_{max} = \in j_{min} \forall\ i \forall k^{min}||x_0\ (k) - x_i\ (k)||$ is the largest value of Δ_{0i};

and x_0 (k) is an ideal sequence.

Calculation of Grey Relational Grade

After averaging the grey relational coefficients, the grey relational grade γ_i can be obtained as

$$\gamma_i = \sum_{k=1}^{n} w_i \xi_i (k) \tag{6}$$

Where n = number of process responses.

From the grey relational grade, the optimal parameters combination is selected. The higher value of the grey relational grade means that the corresponding input parameter is optimal.

ANALYSIS OF VARIANCE (ANOVA)

The role of ANOVA is to explore which factor considerably influences the performance characteristic. This is attained by splitting the total variability of the G.R.G., which is measured by the total of the squared deviations from the total mean of G.R.G. into influences by each factor and the error. First, the total sum of squared deviations S.S.T. from the total mean of grey relational grade can be calculated using equation (7).

$$SS_T = \sum_{i=1}^{n} (Y_L - \overline{Y_{Li}})^2 \tag{7}$$

Where,

n is the number of experiments

Y.L. is the mean of G.R.G. for Lth experiment

The overall sum of the squared deviations S.S.T. is divided into two groups. One is the sum of the squared deviations S.S.M. due to each factor and the sum of the squared error S.S.E. The percentage of influence by each factor in the total sum of the squared deviations S.S.T. can be assessed the significance of the factor in changing the performance characteristic. Finally, the percentage of contribution of each design parameter is given by the following equation (8).

$$\% \text{ of Contribution} = SSM/SST \tag{8}$$

EXPERIMENTAL WORK

Fabrication Process

The Jute fiber sheets of different thicknesses were cut into necessary dimensions as per the mold size and kept ready for the fabrication process. The unsaturated polyester resin was used in the process. We have used Cobalt Naphthenate as a Catalyst to reduce the curing time. To accelerate the catalyst work, a special accelerator called MEKP (Methyl Ethyl Ketone Peroxide) was used as an accelerator. As a result, the curing time is reduced to 4 hours for complete potting. The unsaturated resin, accelerator, and catalyst are bought from M.N. Polyester India Pvt. Ltd. Coimbatore, Tamil Nadu, India. Jute fibers of various thicknesses were collected from the local market. The filler material, copper slag, and Montmorillonite K10 were added to the resin as per the pre-defined composition and stirred using the mechanical stirrer, 1.5 ml of Cobalt Naphthenate was added to the resin as a catalyst, and 1.5 ml of MEKP was added to the resin. The mixture was continuously stirred using a mechanical stirrer while adding the additives.

The L9 orthogonal array is used in determining the most effective compositions. The weight of filler material is calculated for the combinations of 3 different thickness fibers. To calculate weight percentage, sample specimens were prepared using the reinforcement and the resin alone without adding the filler materials.

Finally, all 9 specimens were fabricated (Fig. **1**) using the compression molding technique and were made to cure for 4 hours, as mentioned before. Then, the specimens were cut as per the ASTM standards for wear tests. The composite plates were removed from the mold and cut into various dimensions as per ASTM G 99. As per ASTM standard, the dimension for wear testing specimen is 10 mm × 10 mm.

Wear Test

Wear is associated with interfaces between surfaces, specifically the removal and distortion of material on a surface because of friction action of the opposite surface. The wear of metals happens by the plastic displacement of surface and near-surface material and the detachment of particles that produce wear debris. The size of the created particles may vary from a millimeter to an ion range. This process may be accomplished by interaction with other metals or materials.

(a) Mold

(b) Fabrication process

(c) Compression Moulding

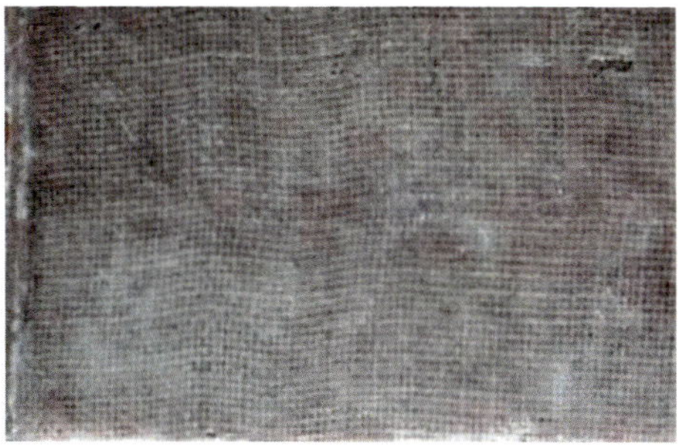

(d) Specimen

Fig. (1). Fabrication of Composite.

In this research, the pin-on-disc (Fig. **2**) testing technique was utilized to test the tribological property of the developed composite material. Initially, the pin surface was kept flat such that it would support the load over its entire cross-section called the first stage. This was achieved by the surfaces of the pin sample ground using emery paper (80 grit size) before testing Run-in-wear was performed in the next stage/ second stage. This phase circumvents the preliminary turbulent period related to friction and wears curves.

Fig. (2). Pin on disc testing machine.

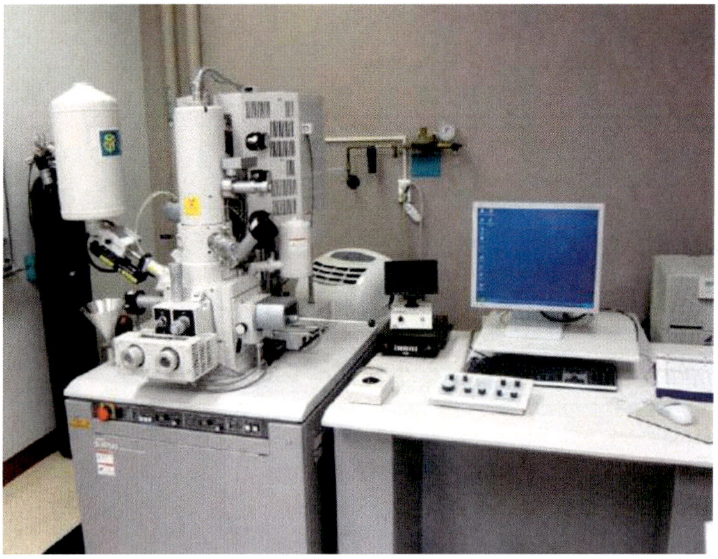

Fig. (3). Scanning Electron Microscope.

Currently, there is a trend to reduce the size of machine elements, increase power, reduce maintenance and improve reliability. These necessities impulse the design limits and require innovative solutions. Testing these solutions on a full machine scale would be too expensive and time-consuming. Therefore, the universal investigational set-ups are regularly utilized to execute tribological research on a prototyping stage, preceding full-scale tests.

Wear can be measured through weight loss or volume loss of the material. Weight loss can also be determined by weighing the material before and after passing through the wear test. The difference in weight of the material gives the material loss due to wearing. The formula to find volume loss is expressed in equation 9.

$$volume\ loss\ (mm^3) = \frac{mass\ loss\ (g)}{density\left(\frac{g}{\ }\right)} * 1000 \quad \textbf{(9)}$$

The coefficient of friction (CoA=πr2F) being a dimensionless number gives the performance of the material during friction. It is scalar and can be denoted as μ. It is the force of friction (F) ratio among 2 contacting faces to the load applied (P).

$$\mu = F/P \quad \textbf{(10)}$$

Pin on disk apparatus is used to conduct tribological studies by varying load and temperature. It consists of a pin held stationary over a disk that rotates at a given speed. Thus, friction develops between the pin and disk, leading to wear on the disk. Then the necessary calculations are made to find the wear loss and volume loss.

Scanning Electron Microscope Test

The investigation of the morphological features of the composites was observed by JEOL M-6390 (Fig. **3**) Scanning Electron Microscope (S.E.M.). It is an essential tool for finding the causes of failure in composite material.

RESULT AND DISCUSSION

Wear Test

First, the composite material is prepared based on the combination of parameters obtained through L9 orthogonal array, and the wear specimen is cut as per the ASTM standard. The wear test was carried out in pin-on-disc testing machine. Special arrangements were included in the machine for our requirement. The disc speed was set to 500rpm, the load was set to 30N, and the time was set to 25min.

For every 5min, the reading was noted for each specimen, and the average result was obtained to study properties.

Table **5** shows the experimental results and the computation of G.R.G. using the procedure. The experimental results show that experiment no 6 produces an optimal wear rate, and experiment 7 produces an optimal result for CoF. G.R.A. is used and recorded in the same table to avoid this confusion. From this table, it is concluded that experiment no. 7 produces better results but only 9 combinations. So it is mandatory to check the others for optimality. For this, the main effects of G.R.G. will be determined (Table **6** & Fig. (**4**)).

Table 5. G.R.A. Calculations.

Ex. No	Wear	COF	S/N Ratio		Normalization		Grey Coefficient		G.R.G.
			Wear	C.O.F.	Wear	C.O.F.	Wear	C.O.F.	
1	1.428	0.33	-3.095	-9.630	0.557	0.000	0.530	0.333	0.432
2	1.182	0.492	-1.452	-6.161	0.467	0.708	0.484	0.631	0.558
3	1.131	0.346	-1.069	-9.218	0.446	0.084	0.474	0.353	0.414
4	0.444	0.532	7.052	-5.482	0.000	0.847	0.333	0.765	0.549
5	1.999	0.384	-6.016	-8.313	0.717	0.269	0.639	0.406	0.522
6	0.922	0.46	0.705	-6.745	0.348	0.589	0.434	0.549	0.491
7	3.157	0.58	-9.985	-4.731	0.935	1.000	0.884	1.000	0.942
8	2.076	0.458	-6.345	-6.783	0.735	0.581	0.654	0.544	0.599
9	3.621	0.468	-11.177	-6.595	1.000	0.620	1.000	0.568	0.784

Table 6. Main effect of G.R.G.

Main effects table for wear rate					
-	1	2	3	Result	Max-Min
Fiber	0.4677	0.5211	0.7750	0.7750	0.3074
Cu	0.6411	0.5596	0.5630	0.6411	0.0815
NC	0.5073	0.6303	0.6260	0.6303	0.1230

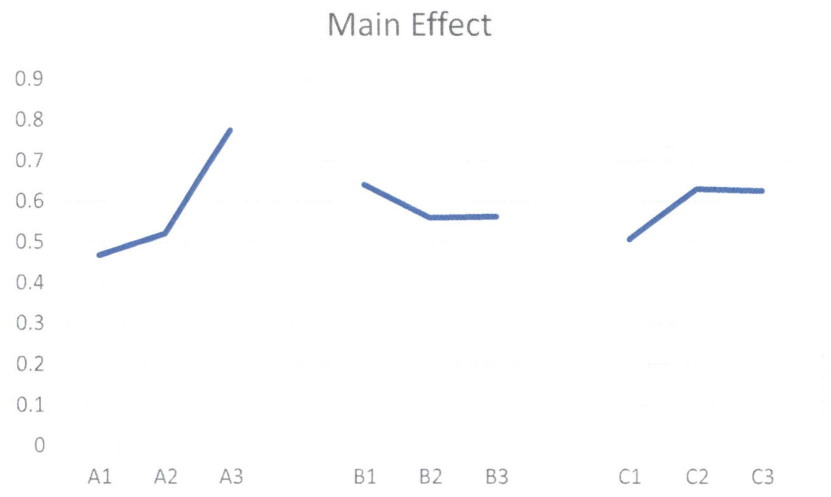

Fig. (4). Main Effects of GRG.

From Table **6**, the optimal combination arrives at levels 3, 1, 2 for fiber, Copper slag % and 2 for nanoclay percentage, respectively. This means the composite material produced by jute and coir fiber, 5% of copper slag, and 2.5% of nanoclay are the optimal combinations. To confirm this combination, a confirmation test is performed on the specimen produced with the optimal combination. The wear rate and CoF are recorded as 1.025 and 0.614 in the confirmation test.

Surface Morphology

Fig. (**5**) shows S.E.M. micrographs of the worn surface of 4 samples (Experiment No 7, 8, 9 & confirmation test) tested under dry sliding friction conditions. Examining the figures, some scratches and small pores are observed on the surfaces. The wear mechanism is thought to be predominantly ploughing and adhesive wear, caused by filler matrix debonding during the wear testing. From this, the confirmation test samples (d) have a lower wear rate as compared to other combinations (a-c). The reason because the surface of sample (d) enhances the property of both the material and epoxy; by this combined combination ratio, the lower wear rate is obtained.

(a) Fiber + Copper Slag (Ex. No; 8)

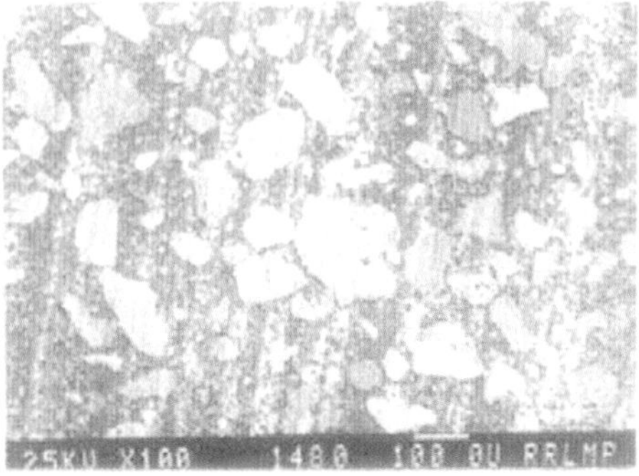

(b) Fiber + Nanoclay (Ex. No; 9)

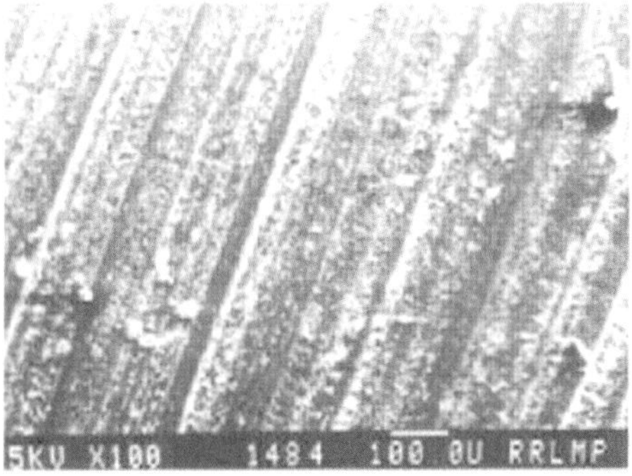

(c) Optimal Condition (Ex. No; 7)

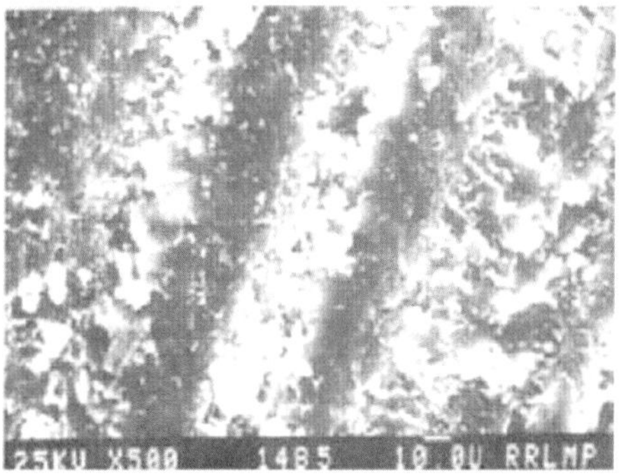

(d) Confirmation Test

Fig. (5). S.E.M. micrographs of the polymer composites.

Determination of Influencing Factor

The influencing criteria among the input parameters are determined using ANOVA method. Table **7** depicts the ANOVA calculations of this work.

Table 7. depicts the ANOVA calculations

FACTOR	DOF	SOS	Mean	Contribution	Percentage
Fiber	2	0.161809587	0.080904793	0.568653	56.87
Cu	2	0.0935	0.04675	0.328590	32.86
NC	2	0.029239481	0.014619741	0.102757	10.28
Total	6	0.284549068	0.142274534	1.000000	100.00

The percentage of influence of the parameters is presented in Fig. (**6**). From Table **7** & Fig. (**6**), it is concluded that the fiber contributes more (56.87%) followed by the percentage of copper slag (32.86%) and nanoclay has the least influence on the final result.

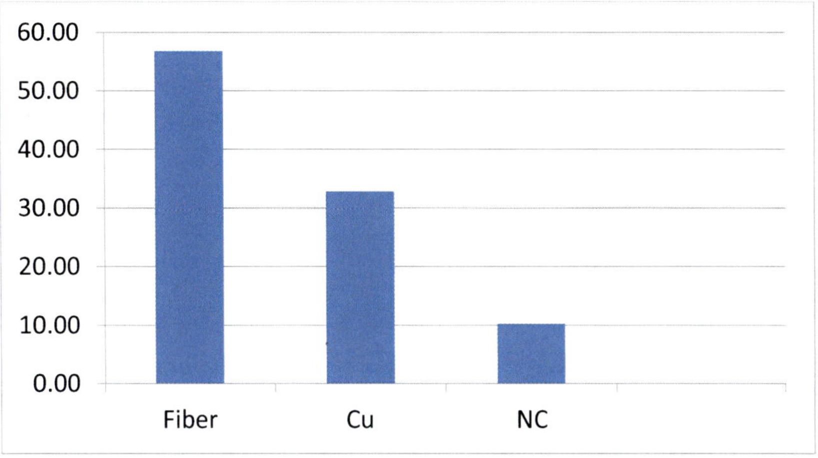

Fig. (6). Percentage of Influence of Parameters.

CONCLUSION

Jute /coir reinforced and copper slag/nanoclay filled polyester composites can be successfully produced by usual compression molding methods as per Taguchi's design of experiment L9 array. In this work, the optimization was carried out to find out the best combination of composite based on the tribological properties. The evaluated result shows that hybrid jute and coir fiber with 5% copper slag and 2.5% nanoclay combination yield good tribological characteristics. Among the three input parameters, fiber influences the tribological properties more than the percentage of copper slag and nanoclay. The results prove that natural fiber-reinforced composites with copper slag & nanoclay fillers enhance tribological properties.

CONSENT FOR PUBLICATION

Not applicable.

CONFLICT OF INTEREST

The author declares no conflict of interest, financial or otherwise.

ACKNOWLEDGEMENTS

Declared none.

REFERENCES

[1] T. Munikenche Gowda, A.C.B. Naidu, and R. Chhaya, "Some mechanical properties of untreated jute fiber reinforced polyester composites", *Compos., Part A Appl. Sci. Manuf.,* vol. 30, no. 3, pp. 277-284, 1999. [http://dx.doi.org/10.1016/S1359-835X(98)00157-2]

[2] M.J. John, and S. Thomas, "Biofibers and biocomposites", *Carbohydr. Polym.,* vol. 71, no. 3, pp. 343-364, 2008. [http://dx.doi.org/10.1016/j.carbpol.2007.05.040]

[3] S. Siddika, F. Mansura, and M. Hasan, "Physico-mechanical properties of jute-coir fiber reinforced hybrid polypropylene composites", *Engineering and Technology,* vol. 73, pp. 1145-1149, 2013.

[4] H. Brodowsky, and E. Mäder, "Jute fibre/epoxy composites: Surface properties and interfacial adhesion", *Compos. Sci. Technol.,* vol. 72, no. 10, pp. 1160-1166, 2012. [http://dx.doi.org/10.1016/j.compscitech.2012.03.025]

[5] W. Hu, Q. Nie, B. Huang, X. Shu, and Q. He, "Mechanical and microstructural characterization of geopolymers derived from red mud and fly ashes", *J. Clean. Prod.,* vol. 186, pp. 799-806, 2018. [http://dx.doi.org/10.1016/j.jclepro.2018.03.086]

[6] M.H. Hasan, M.S. Mollik, and M.M. Rashid, "Effect of nanoclay on thermal behavior of jute reinforced composite", *The International Journal of Advanced Manufacturing Technology,* vol. 94, no. 5, pp. 1863-1871, 2018.

[7] P.S. Mukherjee, and K.G. Satyanarayana, "Structure and properties of some vegetable fibres", *J. Mater. Sci.,* vol. 19, no. 12, pp. 3925-3934, 1984. [http://dx.doi.org/10.1007/BF00980755]

[8] K.G. Satyanarayana, K. Sukumaran, A.G. Kulkarni, S.G.K. Pillai, and P.K. Rohatgi, "Fabrication and properties of natural fiber- reinforced polyester composites", *Composites,* vol. 17, no. 4, pp. 329-333, 1986. [http://dx.doi.org/10.1016/0010-4361(86)90750-0]

[9] Beaumont, W.R. Peter, H.Z. Carl, G. Emmanuēl, T. Ramesh, P. Anoush, W.C. Trevor, B. Marina, and W. Ruggles, *Comprehensive composite materials II.* Elsevier: Amsterdam, The Netherlands, 2018.

[10] J. Rout, M. Misra, S.S. Tripathy, S.K. Nayak, and A.K. Mohanty, "The influence of fiber treatment on the performance of coir–polyester composites", *Compos. Sci. Technol.,* vol. 61, no. 9, pp. 1303-1310, 2001. [http://dx.doi.org/10.1016/S0266-3538(01)00021-5]

[11] H.P.G. Santafe Junior, F.P.D. Lopes, L.L. Costa, and S.N. Monteiro, "Mechanical properties of tensile tested coir fiber reinforced polyester composites", *Revista Matéria,* vol. 15, no. 2, pp. 113-118, 2010.

[12] K. Ramanaiah, A.V. Ratn Prasad, and K. Hema Chandra Reddy, "Effect of fiber loading on mechanical properties of Borassus seed shoot Fiber reinforced polyester composites", *Journal of Materials and Environmental Science,* vol. 3, no. 3, pp. 374-378, 2012.

[13] S.N. Monteiro, L.A.H. Terrones, and J.R.M. D'almeida, "Mechanical performance of coir fiber/polyester composites", *Polym. Test.,* vol. 27, no. 5, pp. 591-595, 2008. [http://dx.doi.org/10.1016/j.polymertesting.2008.03.003]

[14] S. Jayabal, and U. Natarajan, "Influence of fiber parameters on tensile, flexural, and impact properties of nonwoven coir–polyester composites", *Int. J. Adv. Manuf. Technol.,* vol. 54, no. 5, pp. 639-648, 2011.

[http://dx.doi.org/10.1007/s00170-010-2969-8]

[15] R. Messadi, T. Hassine, H. Daly, and F. Lacrampe, "Contribution for a better characterization of the tensile mechanical behaviour of flax/P.P. bio composite materials", *Mater. Res. Express,* vol. 5, no. 12, p. 125504, 2018.
[http://dx.doi.org/10.1088/2053-1591/aae112]

[16] A. Gopinath, M.S. Kumar, and A. Elayaperumal, "Senthil Kumar M and Elayaperumal A., (20145) "Experimental investigations on mechanical properties of jute fiber reinforced composites with polyester and epoxy resin matrices", *Procedia Eng.,* vol. 97, pp. 2052-2063, 2014.
[http://dx.doi.org/10.1016/j.proeng.2014.12.448]

[17] F. Omrani, P. Wang, D. Soulat, M. Ferreira, and P. Ouagne, "Analysis of the deformability of flax-fiber nonwoven fabrics during manufacturing", *Compos., Part B Eng.,* vol. 116, pp. 471-485, 2017.
[http://dx.doi.org/10.1016/j.compositesb.2016.11.003]

[18] R.A. R.A., and Magalhaes P.A.A., "Analysis of the mechanical and thermal properties of jute and glass fiber as reinforcement epoxy hybrid composites", *Materials Science and Engineering,* vol. C 56, pp. 269-273, 2015.
[http://dx.doi.org/10.1016/j.msec.2015.06.031]

[19] H.N. Dhakal, Z.Y. Zhang, R. Guthrie, J. Macmullen, and N. Bennett, "Development of flax/carbon fibre hybrid composites for enhanced properties", *Carbohydr. Polym.,* vol. 96, no. 1, pp. 1-8, 2013.
[http://dx.doi.org/10.1016/j.carbpol.2013.03.074] [PMID: 23688447]

[20] E. Nisini, C. Santulli, and A. Liverani, "Mechanical and impact characterization of hybrid composite laminates with carbon, basalt and flax fibres", *Compos., Part B Eng.,* vol. 127, pp. 92-99, 2017.
[http://dx.doi.org/10.1016/j.compositesb.2016.06.071]

[21] C. Fragassa, A. Pavlovic, and C. Santulli, "Mechanical and impact characterisation of flax and basalt fibre vinylester composites and their hybrids", *Compos., Part B Eng.,* vol. 137, pp. 247-259, 2018.
[http://dx.doi.org/10.1016/j.compositesb.2017.01.004]

[22] R.B. Yusoff, H. Takagi, and A.N. Nakagaito, "Tensile and flexural properties of polylactic acid-based hybrid green composites reinforced by kenaf, bamboo and coir fibers", *Ind. Crops Prod.,* vol. 94, pp. 562-573, 2016.
[http://dx.doi.org/10.1016/j.indcrop.2016.09.017]

[23] N. Venkateshwaran, A. Elayaperumal, and G.K. Sathiya, "Prediction of tensile properties of hybrid-natural fiber composites", *Compos., Part B Eng.,* vol. 43, no. 2, pp. 793-796, 2012.
[http://dx.doi.org/10.1016/j.compositesb.2011.08.023]

[24] M. Jawaid, H.A. Khalil, A. Hassan, R. Dungani, and A. Hadiyane, "Effect of jute fibre loading on tensile and dynamic mechanical properties of oil palm epoxy composites", *Compos., Part B Eng.,* vol. 45, no. 1, pp. 619-624, 2013.
[http://dx.doi.org/10.1016/j.compositesb.2012.04.068]

[25] E. Venkata Subba Reddy, A. Varada Rajulu, K. Hemachandra Reddy, and G. Ramachandra Reddy, "Chemical resistance and tensile properties of glass and bamboo fibers reinforced polyester hybrid composites", *J. Reinf. Plast. Compos.,* vol. 29, no. 14, pp. 2119-2123, 2010.
[http://dx.doi.org/10.1177/0731684409349520]

[26] M.S. Sreekala, J. George, M.G. Kumaran, and S. Thomas, "The mechanical performance of hybrid phenol-formaldehyde-based composites reinforced with glass and oil palm fibres", *Compos. Sci. Technol.,* vol. 62, no. 3, pp. 339-353, 2002.
[http://dx.doi.org/10.1016/S0266-3538(01)00219-6]

[27] A. Hariharan, and H.A. Khalil, "Lignocellulose-based hybrid bilayer laminate composite: Part I-studies on tensile and impact behavior of oil palm fiber-glass fiber-reinforced epoxy resin", *J. Compos. Mater.,* vol. 39, no. 8, pp. 663-684, 2005.
[http://dx.doi.org/10.1177/0021998305047267]

[28] R.M. Nasir, and N.M. Ghazali, "Tribological performance of paddy straw reinforced polypropylene (PSRP) and unidirectional glasspultruded- kenaf (UGPK) composites", *J. Tribol.,* vol. 1, pp. 1-17, 2014.

[29] X. Pei, and K. Friedrich, "Erosive wear properties of unidirectional carbon fiber reinforced PEEK composites", *Tribol. Int.,* vol. 55, pp. 135-140, 2012. [http://dx.doi.org/10.1016/j.triboint.2012.05.020]

[30] M.K. Gupta, and R.K. Srivastava, "Tribological and dynamic mechanical analysis of epoxy based hybrid sisal/jute composite", *Indian J. Eng. Mater. Sci.,* vol. 23, no. 1, pp. 37-44, 2016.

[31] F. Omrani, P. Wang, D. Soulat, and M. Ferreira, "Mechanical properties of flaxfiber- reinforced preforms and composites: Influence of the type of yarns on multi-scale characterizations", *Compos., Part A Appl. Sci. Manuf.,* vol. 93, pp. 72-81, 2017. [http://dx.doi.org/10.1016/j.compositesa.2016.11.013]

[32] A. Abdelbary, and M.N.A. Nasr, "Fatigue wears of polyamides with surface defects under different loading conditions", *Adv. Mat. Res.,* vol. 5, no. 3, pp. 193-203, 2016. [http://dx.doi.org/10.12989/amr.2016.5.3.193]

[33] B. Cheng, and M. Kortschot, "A study of the friction coefficients of unidirectional and woven carbon fibre/epoxy composites", *Polym. Polymer Compos.,* vol. 24, no. 4, pp. 255-263, 2015. [http://dx.doi.org/10.1177/096739111602400404]

[34] N. Chand, and U. Dwivedi, "Effect of coupling agent on abrasive wear behaviour of chopped jute fibre-reinforced polypropylene composites", *Wear,* vol. 261, no. 10, pp. 1057-1063, 2006. [http://dx.doi.org/10.1016/j.wear.2006.01.039]

[35] N. El-Tayeb, "A study on the potential of sugarcane fibers/polyester composite for tribological applications", *Wear,* vol. 265, no. 1-2, pp. 223-235, 2008. [http://dx.doi.org/10.1016/j.wear.2007.10.006]

[36] P. Deepak, H. Sivaraman, R. Vimal, S. Badrinarayanan, and R.V. Kumar, "Study of wear properties of jute/banana fibres reinforced molybdenum disulphide modified epoxy composites", *Mater. Today Proc.,* vol. 4, no. 2, pp. 2910-2919, 2017. [http://dx.doi.org/10.1016/j.matpr.2017.02.172]

[37] D. Wilson, L. Rodic, P. Modak, R. Soos, A. Carpintero, K. Velis, M. Iyer, and O. Simonett, *Global Waste Management Outlook.* United Nations Environment Programme, 2015.

[38] N. Rajini, J.T.W. Jappes, S. Rajakarunakaran, and P. Jeyaraj, "Mechanical and free vibration properties of montmorillonite clay dispersed with naturally woven coconut sheath composite", *J. Reinf. Plast. Compos.,* vol. 31, no. 20, pp. 1364-1376, 2012. [http://dx.doi.org/10.1177/0731684412455259]

[39] N. Saba, M. Jawaid, O.Y. Alothman, M. Paridah, and A. Hassan, "Recent advances in epoxy resin, natural fiber-reinforced epoxy composites and their applications'", *Constr. Build. Mater.,* vol. 106, pp. 149-159, 2016. [http://dx.doi.org/10.1016/j.conbuildmat.2015.12.075]

[40] V.A. Prabu, R.D.J. Johnson, P. Amuthakkannan, and V. Manikandan, "Usage of industrial wastes as particulate composite for environment management: Hardness, tensile and impact studies", *J. Environ. Chem. Eng.,* vol. 5, no. 1, pp. 1289-1301, 2017. [http://dx.doi.org/10.1016/j.jece.2017.02.007]

[41] S. Rajesh, S. Rajakarunakaran, and R.S. Pandian, "Modeling and optimization of sliding specific wear and coefficient of friction of aluminum based red mud metal matrix composite using Taguchi method and response surface methodology", *Materials Physics and Mechanics,* vol. 15, no. 2, pp. 150-166, 2012.

[42] R.X. Liu, and C.S. Poon, "Utilization of red mud derived from bauxite in self-compacting concrete", *J. Clean. Prod.,* vol. 112, pp. 384-391, 2016.

[http://dx.doi.org/10.1016/j.jclepro.2015.09.049]

[43] N. Ye, J. Yang, S. Liang, Y. Hu, J. Hu, B. Xiao, and Q. Huang, "Synthesis and strength optimization of one-part geopolymer based on red mud", *Constr. Build. Mater.*, vol. 111, pp. 317-325, 2016. [http://dx.doi.org/10.1016/j.conbuildmat.2016.02.099]

[44] V. Arumugaprabu, M. Uthayakumar, V. Manikandan, N. Rajini, and P. Jeyaraj, "Influence of red mud on the mechanical, damping and chemical resistance properties of banana/polyester hybrid composites", *Mater. Des.*, vol. 64, pp. 270-279, 2014. [http://dx.doi.org/10.1016/j.matdes.2014.07.020]

[45] S. Biswas, and A. Satapathy, "Tribo-performance analysis of red mud filled glass-epoxy composites using Taguchi experimental design", *Mater. Des.*, vol. 30, no. 8, pp. 2841-2853, 2009. [http://dx.doi.org/10.1016/j.matdes.2009.01.018]

[46] S. Saravanan, and P. Pitchipoo, "Optimization of GMAW parameters to improve the mechanical properties", *Appl. Mech. Mater.*, vol. 813-814, pp. 456-461, 2015. [http://dx.doi.org/10.4028/www.scientific.net/AMM.813-814.456]

[47] I. Siva, J.T.W. Jappes, P. Pitchipoo, S.C. Amico, E.R. Nagarajan, and N. Azhagesan, "Grey optimization of process parameters of surface modification of novel coconut sheath reinforced polymer composites", *Journal of Polymer Engineering,* vol. 33, no. 7, pp. 665-672, 2013. [http://dx.doi.org/10.1515/polyeng-2012-0151]

[48] J.L. Deng, "Introduction to grey system", *J. Grey Syst.,* vol. 1, no. 1, pp. 1-24, 1989.

CHAPTER 8

Application of Natural Fiber-reinforced Composites in Additive Manufacturing

Muthukumaran Ramasamy[1,*], Ajith Arul Daniel[2] and M. Nithya[3]

[1] *CMR Institute of Technology, Hyderabad, Telangana 501401, India*

[2] *Vels Institute of Science, Technology & Advanced Studies, Chennai 600117, Tamil Nadu, India*

[3] *Kakinada Institute of Technological Sciences, Kakinada 533260, Andhra Pradesh, India*

Abstract: The composite materials find their application in almost all industries, *viz.* automobile, aircraft, spacecraft, construction. Polymer composites replace traditional metal matrix composites due to their light weight, high stiffness, and strength with respect to weight. Extensive research is being carried out by introducing reinforcing fibers in the polymer matrix to make the composite more functional, versatile, and on par with the metal composites. The increase in environmental awareness towards sustainability has attracted researchers to utilize natural fibers in polymer composites. The conventional way of preparing fiber-reinforced polymer composites are compression moulding, injection moulding, *etc.* With the progression of industry 4.0, the manufacturing industry has been shifted towards additive manufacturing, which serves as one of the pillars of industry 4.0. In this context, the manufacturing of polymer composites with natural fibers through additive manufacturing is gaining importance to provide sustainable development. This chapter integrates the review of existing literature to categorize various additive manufacturing technologies being adopted for natural fiber reinforced polymer composites (NFRPC). It also critically analyses the challenges and opportunities associated with additive manufacturing of NFRPC.

Keywords: Additive manufacturing, Composites, Natural fiber, 3D printing.

INTRODUCTION

The intense environmental concern over the unsustainable intake of petroleum products has led to adopting environment-friendly materials. Several manufacturing industries and the scientific research community have started adopting the key sustainability principles such as reduce, recycle, repurpose, recover, replace, redesign, remanufacture and finally refuse [1]. The endurance of

* **Corresponding author Muthukumaran Ramasamy:** CMR Institute of Technology, Hyderabad, Telangana 501401, India; E-mail: muthufs25@gmail.com

C. Samson Jerold Samuel, M. Suresh & S. Gnanasekaran (Eds.)

any manufacturing industry depends on the various material utilised, and it is considered a pillar for the industry [2]. The manufacturing industry adopts a variety of materials, including metals, alloys, composites, and petroleum products. The increased use of non-renewable resources like coal, petroleum-based fuels and products has threatened the manufacturing industry. At the current consumption rate, petrol may last only for another 50 – 60 years [3, 4]. The shortfall of pure metals to meet the contemporary product descriptions led to the development of composite materials to manufacture novel products. With the growing interest and awareness among the end-users towards adopting environmentally friendly and sustainable products, manufacturers have also shifted their focus to sustainable materials. The current research fraternity focuses on natural fiber reinforced polymer composites for the replacement of synthetic fibers as reinforcement and desirable properties at an affordable cost [5, 21].

Natural Fibers

The natural fibers can generally be grouped into two types *viz.*, naturally occurring inorganic fibers and organic fibers, including plant or animal-based fibers [13]. Plant fibers have commercial and technical importance and have been used by humankind for over 40,000 years [6 - 8]. Many engineering designs have been created with the inspiration of plants and their products through biomimicry.The plants that have grown enormously in tropical regions are being utilized for various applications such as ropes, beds, and bags [9 - 11]. Based on the location of fibers within the plant, fibers can be classified into primary, secondary, leaf, bast and seed fibers, as shown in Fig. (**1**). Fig. (**2**) shows different types of plant fiber commonly used as reinforcement in the composite material.

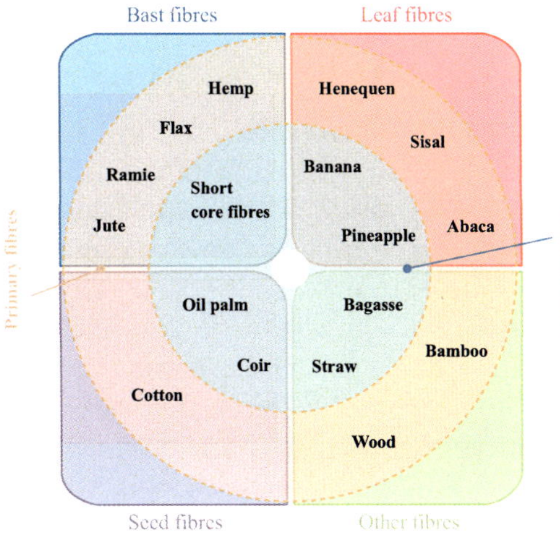

Fig. (1). Classification of plant fibers [12, 30].

Fig. (2). Plant fibers [16].

The potential application of natural fiber-reinforced composites includes the Automotive industry (door panels, seat backs, dashboard, spare wheel), aircraft industry (interior panelling), construction industry (railing, bridge), household products and furniture, electrical products (mobile, laptop cases), sports items (bicycle frames, snowboards), packaging industry and biomedical sector [18 - 20, 39]. In addition to the cost reduction, the use of natural fibers helps achieve stiffness-to-weight ratio, thermal insulation, CO_2 neutrality, and recyclability [17]. But, natural fibers use is prone to some disadvantages such as thermal instability, moisture absorption, and bonding which restricts their structural application [13, 14]. However, the recent research towards the hybridization of natural and artificial fibers will overcome the disadvantage [15].

Additive Manufacturing

Additive Manufacturing (AM), commonly called 3D printing (3DP), has earned exponential growth and popularity in recent decades because of its design freedom, flexibility in fabrication, material savings, and ability to print complex customized geometry with high precision and accuracy [22, 25, 35]. 3D printing technology also has the potential to print a broad spectrum of materials (metals, polymers– thermoplastic/thermosetting, ceramics, concrete) with the least manufacturing time and cost of production [26, 32]. Moreover, the above-said advantages can be achieved without any tooling process [23]. In this modern era, 3D Printing finds its application in almost all industries not limited to aerospace, automotive, construction, bio-medical, fashion, and jewellery [24, 36 - 38]. Composite material fabrication has been revolutionized with the advent of 3D printing/AM technology [25]. Polymer composite with the addition of natural fibers as reinforcement has recently been recognised for its potential application in additive manufacturing. Natural Fiber Reinforced Polymer Composite

(NFRPC) possess desired mechanical properties such as tensile strength, flexural strength, tensile and elastic modulus, and high strength–to–weight ratio than conventional materials and hence can be opted to replace metallic materials [27]. Over the past few years, various additive manufacturing techniques for composite materials have been developed. Each method has its pros and cons, and its application depends on the type of material used for printing [28, 33, 34]. Fiber-reinforced Polymer composite (FRPC) additive manufacturing techniques can be grouped under two categories *viz*., Long fiber-reinforced polymer composite and short fiber-reinforced polymer composite manufacturing [29, 31]. Fig. (**3A**) and Fig. (**3B**) show the different primary additive manufacturing techniques available for long and short FRPC.

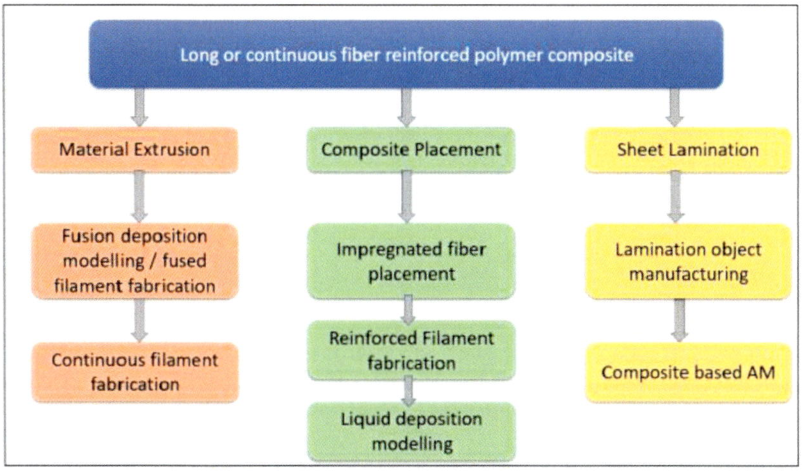

Fig. (3A). Additive manufacturing technique for long FRPC.

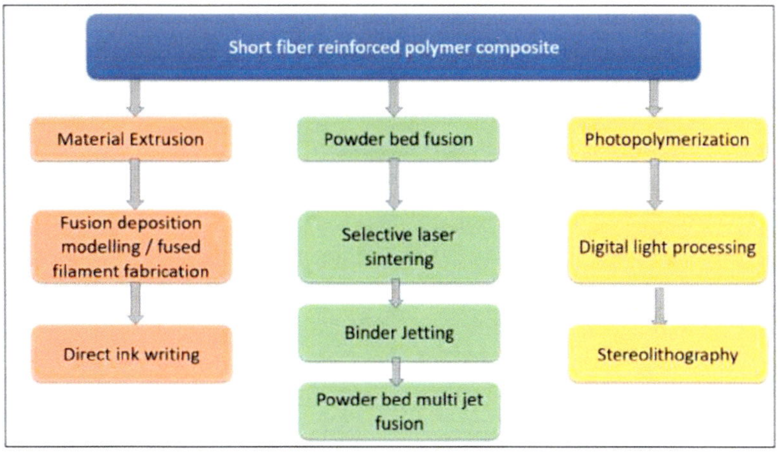

Fig. (3B). Additive manufacturing technique for short FRPC.

The basic needs (sustainability and environment friendly) coupled with advanced manufacturing technique have led to plant fibers in Additive manufacturing. For better understanding, the state-of-the-art additive manufacturing of natural fiber-reinforced composites is being discussed in the following section.

ADDITIVE MANUFACTURING OF NFRPC

David Stoof *et al.* investigated the factors affecting fusion deposition modelling of polylactic acid (PLA) polymer composite components reinforced with natural fibers. Two different types of fibers, such as hemp and harakeke fiber, with a uniform size of 3mm, were added into the polymer matrix with varying percentages such as 10, 20, and 30%. The fibers were treated through a Castin granulator using 5% NaOH as a treatment solution for hemp fiber and 5% NaOH along with 2% sodium sulphite for harakeke fiber. The fibers exhibit reduction in diameter after treatment. Composite specimens with varying fiber content were mixed in a heat chamber having rotating shafts with an average speed of 16 rpm at 185°C and allowed to cool room temperature. The prepared composite undergoes granulation *via* 8 mm sieve and is extruded with 3 mm diameter die. The authors have developed an electric spooling equipment to consistent filament cross-section. Five dog bone samples (of size 10 x 15 x 3 mm) for each percentage of fiber content were printed using a 3D printer (2014 Diamond age) from the sealed filament with a heating bed temperature at 110°C. From the microstructure study, composites printed through 1 mm die have semi-blockages for more than 20% fiber content. The same effect can be eliminated with 2 mm extrusion but decreased definition. The increase in fiber content beyond 20% wt. affects the microstructure and reduces the tensile strength. The tensile strength and Young's modulus values of 20% harakeke fiber samples are found to have higher values than plain PLA samples by 42.29% and 5.4%, respectively.

Vamsi Krishna Balla *et al.* adopted the fused filament fabrication (FFF) 3D printing technique to produce thermoplastic copolyester (TPC) composites reinforced with soya bean-derived hull fibers. The soya bean hull fibers were treated with dilute acid hydrolysis. Soya bean hull fibers (both hydrolyzed and unhydrolyzed) were used to fabricate 5 and 10 wt.% TPC. The fibers were subjected to different types of chemical treatment such as dry blended, wet shear mixed, single-stage, and two-stage hydrolyzed hull as-received hulls. Composite specimens with varying fiber content were mixed using a torque rheometer with a speed of 50 rpm at 180°C for 18 to 20 min. Small composite granules of size varying between 1 to 5mm were extruded into 1.75 mm diameter filaments at 165°C using a tungsten carbide die. The microstructure study revealed that composites made using treated (dilute acid hydrolysis) fibers exhibit reduced hydrophilicity sound interface bonding with no visible gap. The 3D printed composite samples were subjected to a tensile test to determine the strengthening effect of tensile stress at 5% and 50%. The results have revealed that 5% wt. treated fibers retained the strengthening effect and recorded elastic modulus between 54 MPa and 56 MPa, which is 36% more than the elastic modulus of pure TPC. However, authors have found that 10% wt of 2-stage hydrolyzed fibers show improvement in the elastic modulus value up to 54% and strain up to 38% for stress at 50%. They also concluded that soya bean hull fibers with appropriate chemical treatment could be used as a potential reinforcement in polymer composites using additive manufacturing.

A.C. Araujo *et al.* proposed a composite comprising gypsum sisal fiber using binder jetting (BJ). Gypsum powder was the matrix with bimodal particle distribution and a surface area of 0.866 m^2/g. Sisal fibers were cut to length 1 – 2 mm and were added to the gypsum matrix at 1.5% by volumetric proportion. A commercially available binder called 'VisiJet PXL Clear' was used as a binder, and ethyl-2-cyanoacrylate was used as an infiltrant. In addition to the mechanical test *i.e.*, flexural strength and porosity, specimens were subjected to porosity, SEM, XRD, and thermogravimetric analysis. The BJ printer Projet460 Plus was used to print the samples with and without fibers having two geometries; cube specimens for thermal and X-ray analyses and prismatic beam for flexural and porosity tests. Design of experiments using ANOVA was performed to identify the factors influencing flexural strength such as fiber addition, length and flexure direction, and infiltration. The XRD and SEM analysis found that composite matrix printed with sisal fibers have pores and a low dihydrate phase presence due to lack of hydration time while depositing the binder. Porosity analysis reveals that the matrix with sisal fibers increases porosity due to a lack of interaction with the gypsum matrix. The addition of fibers has decreased the specimens' strength by 32%. However, the strength of the composites has shown a 41% increase during post-process infiltration.

Yizhou Jiang *et al.* developed a highly transparent composite having natural raw flax fibers as reinforcement through additive manufacturing. Raw flax fibers that are hollow and irregular in cross-section were chopped into 1 to 3 mm length. The fibers were subjected to a serious treatment as follows: immersion in nitric acid for 15 min, washed using in deionized water to eradicate remaining acids, the sonicating process for 30 min and finally immersed in 2.5 mol/L Sodium hydroxide and 0.4 mol/L Sodium Sulphitein deionized water at 90 °C for 6 h and transferred into 2.5 mol/L Hydrogen Peroxide in deionized water at room temperature for 72h to remove the entire lignin. After the rigorous chemical treatment, the fibers were stored in ethanol and finally, they became transparent. A room temperature curing polymer Silicone Elastomer having base and curing agent as two components was used to prepare the composite with varying fiber percentages such as 1%, 2%, and 5% by weight. The treated fibers were mixed with the elastomer base and blended with curing at 2000 rpm for 30 min and 3 min, respectively. The tensile strength and Young's modulus values increased for all specimens with fiber reinforcement compared to the pure elastometer. The printability of the treated flax fiber composite (5 wt% fiber) was tested by fabrication loop pattern using extrusion-based 3D printing. The authors have developed a transparent composite using treated flax fiber with 90% light transmittance tested using a spectrophotometer. It also exhibits superior tensile performance and excellent printability compared to synthetic fiber composites.

A. Le Duigou *et al.*, customized PLA composite material reinforced with continuous flax fiber through fusion deposition modelling. Continuous fiber fabrication process was adopted to produce the filaments through extrusion coating. Microstructure observations using SEM reveal irregularities on the printed filament surface due to the co-extrusion process. The porosity within the continuous flax fiber is low compared to the short natural fiber composite. High-quality printed composites can be achieved by limiting the continuous flax yarn volume content to 35%. Finally, the authors have recommended untwisted flax yarn as the twisted one has prevented full impregnation.

Table 1. Types of Plant fiber utilized in Additive manufacturing and the methods adopted.

Ref	Fiber	Treatment	% of fiber content	Matrix	Tensile strength MPa	Method of printing	Printer name
[41]	Harakeke fiber Hemp fibr	✓	10, 20, 30 (weight)	PLA	0 – 35 Hemp – (10% - 37.5, 20% - 29, 30% - 24) Harakeke – (10% - 34, 20% - 37, 30% - 28.5)	Fused deposition modelling	Diamond age 3D printer
[42]	Soybean hull fiber	✓	5, 10	TPC	-	Fused filament fabrication	Printrbot desktop machine
[43]	Sisal fiber	✗	1.5 (volume)	gypsum	0 – 7.86 1.5 – 12.52	Binder Jetting process	
[44]	Flax fibers	✓	1, 2, 5 (weight)	Silicone Elastomer	0 – 1.75 1 – 2.5 2 – 3.3 5 – 4.2	Fusion deposition modelling (extrusion-based)	E3V, Nordson EFD

(Table 1) cont.....

Ref	Fiber	Treatment	% of fiber content	Matrix	Tensile strength MPa	Method of printing	Printer name
[45]	Continuous Flax fiber	✗	35 (volume)	PLA	-	Fusion deposition modelling (extrusion-based)	Prusa i3 printer
[46]	Silk fiber Wool	✓	Hand lay-up technique	PLA	16.03-24.58	Fused filament deposition	Destock printer
[47]	Carbon fibers, & twisted yarns of natural jute fibers	✗	6.6 (carbon) 6.1 (jute) (volume)	PLA	PLA – 48.5 Carbon – 185.2 Jute – 57.1	Fused-deposition modeling	FDM Blade-1 3D printer
[48]	Natural rock Basalt fiber	✓	5, 10, 20	PLA	0 – 54.2 5 – 63 10 – 66.5 20 – 71.9	Fused-deposition modeling	Ultimaker 2 + printer
[49]	Pine apple fiber	✗	4 (volume)	PLA	0 – 89.44 4 – 96.78	Fused-deposition modeling	ANET A8 3D Printer Prusa I3
[50]	Flax fiber	✗	10.6 20.4 36.7	PLA	0 – 47 10.6 – 55 20.4 – 89 36.7 - 80	Fused filament fabrication	Five-axis 3D printer

✓ - indicates fiber has undergone chemical treatment, ✗ - indicates raw fibers without any treatment

Amarveer Singh Mangat *et al.*, fabricated PLA composite using fused filament deposition and manually inserting (hand-lay method) the waste natural silk fiber and sheep wool fiber for biomedical applications. The fibers were decontaminated using sodium hypochlorite (NaOCl) and cleaned using hydrogen peroxide with a stirring capacity of 10 Hz for 30 min. The fiber was then dried at 700°C for 24 hours in a hot air oven. The authors have faced challenges while inserting the fibers manually and suggested an automatic fiber insertion mechanism. The flexural specimens and laminated structures using layers of matrix fibers were manufactured using fused filament deposition. The process parameters such as density, number of laminates, and type of fiber were optimized through ANOVA. In addition to mechanical properties, specimens were selected randomly for bacterial testing. The silk fiber PLA has more strength and hardness than the sheep wool fiber PLA. However, PLA sheep wool composite is found to have low biodegradation than silk fiber composite.

Ryosuke Matsuzaki *et al.* utilized PLA filament as the matrix for fused-deposition modeling, comprising two different reinforcements *viz*., carbon fiber and twisted jute fiber. Quasistatic tensile tests and scanning electron microscopy tests were conducted on the specimen by orienting the fibers along the longitudinal direction, and hence the mechanical properties revealed were orthotropic in nature. In order to enhance the permeation of fibers into the resin, the fibers are heated to 210°C before entering the nozzle. The volume fractions of reinforcing fibers are found to be 6.6% (carbon) and 6.1% (jute) during the optical examination. 3D printed carbon fiber composites show enhanced properties compared to jute fiber composites. However, the fracture imaging depicts fiber pullout for both types of reinforcement due to insufficient adhesion between fiber and resin. The authors have suggested providing suitable tension and torsion while using a twisted yarn as reinforcement and the fibers need to be treated before the printing process.

Lin Sang *et al.* aimed to fabricate PLA with silane-treated natural rock Basalt fiber through 3D printing and compared the results with 3D printed PLA reinforced with carbon fiber. Basalt and carbon fibers are cut into pieces of length 1, 2, and 3 mm for blending. Basalt fiber was treated with a silane coupling agent to modify the fiber surface for better adhesion with the matrix. The materials, matrix, and fiber were oven-dried at 60°C for a period of 36 hours before blending. The blended and kneaded materials are cooled down to granules and hot-pressed in a compression mould at 200°C for 10 min with a pressure of 5 MPa. Fiber-filled filaments are fabricated through a singer-screw extruder with a filament diameter of 1.75 mm and stored under standard conditions prior to printing. Ultimaker2 + printer was used to print specimens for tensile, impact, and flexural tests from the fabricated filaments. The extrusion process was carried out with a nozzle temperature of 210 °C and a platform temperature of 60 °C. The specimens prepared through compression moulding and 3D printing exhibit similar tensile strength. The mechanical properties and micrographs reveal the superior performance of basalt fiber compared to carbon fiber for the same weight fraction. The bonding defects and poor flowability are reasons for the failure of carbon-reinforced composites. The difference in fiber length is the reason for the improved tensile strength of basalt fiber-reinforced composite.

Jaya Suteja *et al.* investigated the mechanical properties of 3D printed PLA composite reinforced with pineapple leaf fiber. The fiber is extracted from the pineapple leaf by removing the gummy substance. The fibers are cylindrical in a cross section with lengths ranging between 55 to 75 cm and a diameter of 0.5 – 1 mm. The pineapple fiber with a diameter 0.1 mm with the volume ratio of 1:4 is used in PLA matrix to produce a filament diameter of 1.75 mm as per 3D printer requirement. The nozzle temperature varies between 200 °C and 210 °C, keeping other process parameters constant. In total, 18 specimens were printed to test the tensile strength, elongation, and dimensional error due to 3D printing. The average strength of the PLA specimen is lower than the composite specimen as the fiber content in the matrix improves the tensile strength. Similarly, high extrusion temperature and feed rate also increase the tensile strength of the fiber-reinforced composite. However, authors have suggested optimizing the temperature as extrusion temperature increases the elongation of composite parts.

Haiguang Zhang *et al.* developed flax fiber-reinforced plastic composite through the fused filament fabrication method. The PLA solid pellets were heated to 190 °C, and the band was heated up to 300 °C. The molten PLA resin was extruded at a speed of 200 to 600 mm/min, and flax fibers of 0.5 mm diameter were laid and coated by PLA matrix. A five-axis 3D printer was used to print the specimens for the tensile test and curved bending test. The diameter of the filament and building orientation is considered as process parameters. Maximum tensile strength of 89 MPa was achieved for 1 mm flax fiber coated PLA, whereas the PLA without any fiber achieved a tensile strength of 47 MPa. The tensile strength of 0.8 mm flax fiber composite was slightly reduced due to improper coating of PLA resin which is attributed as one of the limitations of the printer. As far as the building orientation is considered, flat printing has recorded a maximum bending load of 119 N, which is lower than flat printing. The reason behind the reduced bending load was filling compactness and void formation. However, curved printing has improved the surface quality and removed the stair-step effect. Further, the study has concluded that natural flax fiber can effectively be used to fabricate functional parts through fused filament fabrication.

FUTURE SCOPE

Industry 4.0 has made the manufacturing industry intelligent and autonomous through additive manufacturing, which demands the use of multifunctional and sustainable materials. The environmental and recycling challenges faced by conventional materials can be fixed through novel materials manufactured by a novel technique, additive manufacturing. Functional composite materials and optimized structures for a wide range of applications can be achieved through parallel development in material research and additive manufacturing. NFPRC composite research has future scope for thermal-related studies as they have not been explored so far.

CONCLUSION

Through this critical review, different aspects discussed were summarized as follows:

- Before printing, suitable chemical treatment (preferably alkali treatment) of plant fiber is essential to enhance the interface bonding with the matrix and reduce the hydrophilic tendency.
- (Polylactic acid) PLA is a polymer matrix widely adopted for additive manufacturing with unique characteristics like biodegradable, recyclable, and eco-friendly.
- The fusion deposition method is the frequently adopted process for 3D printing of NFRPC.
- Minor voids are formed during the printing process, which needs to be researched through SEM analysis.
- The mechanical properties of the NFRPC composites are influenced by the process parameters such as diameter and length of the fiber, nozzle diameter and temperature, fiber content, the orientation of fiber, and manufacturing technique.

Based on the studies performed in this chapter, it is evident that natural fibers reinforced composites can be useful in making complex and functional parts using additive manufacturing. However few challenges faced during the 3D printing of natural fiber reinforced composites are cluster, void formation, nozzle clogging and fiber-matrix bonding. These defects can easily be rectified by properly selecting process parameters chemical treatment of fibers before printing. Finally, basic knowledge of the fiber-matrix interaction and influence of natural fibers on the rheological properties of polymer matrices is necessary. This chapter serves as a guide for those who start their research towards additive manufacturing polymer composite reinforced with natural fibers.

CONSENT FOR PUBLICATION

Not applicable.

CONFLICT OF INTEREST

The author declares no conflict of interest, financial or otherwise.

ACKNOWLEDGEMENTS

Declared none.

REFERENCES

[1] V.K. Balla, K.H. Kate, J. Satyavolu, P. Singh, and J.G.D. Tadimeti, "Additive manufacturing of natural fiber reinforced polymer composites: Processing and prospects", *Compos., Part B Eng.,* vol. 174, p. 106956, 2019.
[http://dx.doi.org/10.1016/j.compositesb.2019.106956]

[2] L. Kerni, S. Singh, A. Patnaik, and N. Kumar, "A review on natural fiber reinforced composites", *Mater. Today Proc.,* vol. 28, pp. 1616-1621, 2020.

[http://dx.doi.org/10.1016/j.matpr.2020.04.851]

[3] M.P. Ho, "Critical factors on manufacturing processes of natural fiber composites", *Compos., Part B Eng.,* vol. 43, no. 8, pp. 3549-3562, 2012. [http://dx.doi.org/10.1016/j.compositesb.2011.10.001]

[4] J.D. Bronzino, *Biomedical engineering handbook: biomedical engineering fundamentals.* CRC Press, 1999.

[5] G. Cicala, G. Cristaldi, G. Recca, and A. Latteri, "Composites Based on Natural Fibre Fabrics", In: *Woven Fabric Engineering.,* P. Dubrovski, Ed., InTech: London, UK, 2010. [http://dx.doi.org/10.5772/10465]

[6] A. Bourmaud, J. Beaugrand, D.U. Shah, V. Placet, and C. Baley, "Towards the design of high-performance plant fibre composites", *Prog. Mater. Sci.,* vol. 97, pp. 347-408, 2018. [http://dx.doi.org/10.1016/j.pmatsci.2018.05.005]

[7] E. Kvavadze, O. Bar-Yosef, A. Belfer-Cohen, E. Boaretto, N. Jakeli, Z. Matskevich, and T. Meshveliani, "30,000-year-old wild flax fibers", *Science,* vol. 325, no. 5946, p. 1359, 2009. [http://dx.doi.org/10.1126/science.1175404] [PMID: 19745144]

[8] S. Reichert, A. Menges, and D. Correa, "Meteorosensitive architecture: Biomimetic building skins based on materially embedded and hygroscopically enabled responsiveness", *CAD Comput. Aided Des.,* vol. 60, pp. 50-69, 2015. [http://dx.doi.org/10.1016/j.cad.2014.02.010]

[9] L. Yan, N. Chouw, and K. Jayaraman, "Flax fibre and its composites - A review", *Compos., Part B Eng.,* vol. 56, pp. 296-317, 2014. [http://dx.doi.org/10.1016/j.compositesb.2013.08.014]

[10] C. Alves, "Ecodesign of automotive components making use of natural jute fiber composites", *J. Clean. Prod.,* vol. 18, no. 4, pp. 313-327, 2010. [http://dx.doi.org/10.1016/j.jclepro.2009.10.022]

[11] A.K. Bledzki, V.E. Sperber, and O. Faruk, "Natural and wood fibre reinforcement in polymers. iSmithersRapra Publishing",

[12] D.U. Shah, "Natural fibre composites: Comprehensive Ashby-type materials selection charts", *Mater. Des.,* vol. 62, pp. 21-31, 2014. [http://dx.doi.org/10.1016/j.matdes.2014.05.002]

[13] K.V. Balaji, K. Shirvanimoghaddam, G.S. Rajan, A.V. Ellis, and M. Naebe, "Surface treatment of Basalt fiber for use in automotive composites", *Mater. Today Chem.,* vol. 17, p. 100334, 2020. [http://dx.doi.org/10.1016/j.mtchem.2020.100334]

[14] J. Holbery, and D. Houston, "Natural-fiber-reinforced polymer composites in automotive applications, JOM", *J. Occup. Med.,* vol. 58, pp. 80-86, 2006.

[15] M. Ramasamy, and A.A. Daniel, "M. Nithya, S. Sathees Kumar, R. Pugazhenthi, Characterization of natural – Synthetic fiber reinforced epoxy based composite – Hybridization of kenaf fiber and kevlarfiber", *Mater. Today Proc.,* 2020.

[16] S.R. Benin, S. Kannan, R.J. Bright, and A. Jacob Moses, "A review on mechanical characterization of polymer matrix composites & its effects reinforced with various natural fibres", *Mater. Today Proc.,* 2020.

[17] V.K. Balla, K.H. Kate, J. Satyavolu, P. Singh, and J.G.D. Tadimeti, "Additive manufacturing of natural fiber reinforced polymer composites: Processing and prospects", *Compos., Part B Eng.,* vol. 174, no. March, p. 106956, 2019. [http://dx.doi.org/10.1016/j.compositesb.2019.106956]

[18] M. Ramesh, K. Palanikumar, and K.H. Reddy, "Plant fibre based bio-composites: sustainable and renewable green materials", *Renew. Sustain. Energy Rev.,* vol. 79, pp. 558-584, 2017.

[http://dx.doi.org/10.1016/j.rser.2017.05.094]

[19] O. Faruk, A.K. Bledzki, H.P. Fink, and M. Sain, "Progress report on natural fiber reinforced composites", *Macromol. Mater. Eng.,* vol. 299, no. 1, pp. 9-26, 2014. [http://dx.doi.org/10.1002/mame.201300008]

[20] J. Holbery, and D. Houston, "Natural-fiber-reinforced polymer composites in automotive applications", *JOM,* vol. 58, no. 11, pp. 80-86, 2006. [http://dx.doi.org/10.1007/s11837-006-0234-2]

[21] L. Mohammed, M. N. M. Ansari, G. Pua, M. Jawaid, and M. S. Islam, "A Review on Natural Fiber Reinforced Polymer Composite and Its Applications", *Int. J. Polym. Sci,* 2015. [http://dx.doi.org/10.1155/2015/243947]

[22] G.D. Goh, Y.L. Yap, S. Agarwala, and W.Y. Yeong, "Recent Progress in Additive Manufacturing of Fiber Reinforced Polymer Composite", *Adv. Mater. Technol.,* vol. 4, no. 1, pp. 1-22, 2019. [http://dx.doi.org/10.1002/admt.201800271]

[23] V. Dikshit, G.D. Goh, A.P. Nagalingam, G.L. Goh, and W.Y. Yeong, *"Recent progress in 3D printing of fiber-reinforced composite and nanocomposites," in Fiber-Reinforced Nanocomposites: Fundamentals and Applications.* Elsevier, 2020, pp. 371-394. [http://dx.doi.org/10.1016/B978-0-12-819904-6.00017-7]

[24] B.P. Conner, "Making sense of 3-D printing: creating a map of additive manufacturing products and services", *Addit. Manuf.,* vol. 1, pp. 64-76, 2014. [http://dx.doi.org/10.1016/j.addma.2014.08.005]

[25] B. Berman, "3-D printing: The new industrial revolution", *Bus. Horiz.,* vol. 55, no. 2, pp. 155-162, 2012. [http://dx.doi.org/10.1016/j.bushor.2011.11.003]

[26] G.D. Goh, Y.L. Yap, H.K.J. Tan, S.L. Sing, G.L. Goh, and W.Y. Yeong, "Process-structure-properties in polymer additive manufacturing *via* material extrusion: a review"., *Crit. Rev. Solid State Mater. Sci.,* vol. 8, no. 5, p. 55, 2019.

[27] P.K. Mallick, *Fiber-Reinforced Composites: Materials, Manufacturing, and Design.* CRC Press, 2008.

[28] G.D. Goh, S. Agarwala, G.L. Goh, V. Dikshit, S.L. Sing, and W.Y. Yeong, "Additive manufacturing in unmanned aerial vehicles (UAVs): Challenges and potential", *Aerosp. Sci. Technol.,* vol. 63, pp. 140-151, 2017. [http://dx.doi.org/10.1016/j.ast.2016.12.019]

[29] V. Dikshit, G.D. Goh, A.P. Nagalingam, G.L. Goh, and W.Y. Yeong, *"Recent progress in 3D printing of fiber-reinforced composite and nanocomposites," in Fiber-Reinforced Nanocomposites: Fundamentals and Applications.* Elsevier, 2020, pp. 371-394. [http://dx.doi.org/10.1016/B978-0-12-819904-6.00017-7]

[30] A. Bourmaud, J. Beaugrand, D.U. Shah, V. Placet, and C. Baley, "Towards the design of high-performance plant fibre composites", *Prog. Mater. Sci.,* vol. 97, pp. 347-408, 2018. [http://dx.doi.org/10.1016/j.pmatsci.2018.05.005]

[31] M. Domm, *Printing of three-dimensional polymer composite structures with continuous fiber reinforcement.* Elsevier Inc., 2020. [http://dx.doi.org/10.1016/B978-0-12-819535-2.00011-9]

[32] S. Dritsas, Y. Vijay, M. Dimopoulou, N. Sanadiya, and J.G. Fernandez, *Robotic Fabrication in Architecture, Art and Design.* Springer International Publishing, 2019.

[33] I. Fidan, "The trends and challenges of fiber reinforced additive manufacturing", *Int. J. Adv. Manuf. Technol.,* vol. 102, no. 5–8, pp. 1801-1818, 2019. [http://dx.doi.org/10.1007/s00170-018-03269-7]

[34] G.D. Goh, Y.L. Yap, S. Agarwala, and W.Y. Yeong, "Recent Progress in Additive Manufacturing of

Fiber Reinforced Polymer Composite", *Adv. Mater. Technol.,* vol. 4, no. 1, pp. 1-22, 2019. [http://dx.doi.org/10.1002/admt.201800271]

[35] T. Hofstätter, D.B. Pedersen, G. Tosello, and H.N. Hansen, "State-of-the-art of fiber-reinforced polymers in additive manufacturing technologies", *J. Reinf. Plast. Compos.,* vol. 36, no. 15, pp. 1061-1073, 2017. [http://dx.doi.org/10.1177/0731684417695648]

[36] I. Karakurt, and L. Lin, "3D printing technologies: techniques, materials, and post-processing", *Curr. Opin. Chem. Eng.,* vol. 28, pp. 134-143, 2020. [http://dx.doi.org/10.1016/j.coche.2020.04.001]

[37] J. Liu, L. Sun, W. Xu, Q. Wang, S. Yu, and J. Sun, "Current advances and future perspectives of 3D printing natural-derived biopolymers", *Carbohydr. Polym.,* vol. 207, pp. 297-316, 2019. [http://dx.doi.org/10.1016/j.carbpol.2018.11.077] [PMID: 30600012]

[38] R. Matsuzaki, M. Ueda, M. Namiki, T.K. Jeong, H. Asahara, K. Horiguchi, T. Nakamura, A. Todoroki, and Y. Hirano, "Three-dimensional printing of continuous-fiber composites by in-nozzle impregnation", *Sci. Rep.,* vol. 6, p. 23058, 2016. [http://dx.doi.org/10.1038/srep23058] [PMID: 26965201]

[39] L. Mohammed, M.N.M. Ansari, G. Pua, M. Jawaid, and M.S. Islam, "A Review on Natural Fiber Reinforced Polymer Composite and Its Applications", *Int. J. Polym. Sci.,* vol. 2015, pp. 1-15, 2015. [http://dx.doi.org/10.1155/2015/243947]

[40] V. Sekar, M.H. Fouladi, S.N. Namasivayam, and S. Sivanesan, "Additive Manufacturing: A Novel Method for Developing an Acoustic Panel Made of Natural Fiber-Reinforced Composites with Enhanced Mechanical and Acoustical Properties", *J. Eng. (United Kingdom),* vol. 2019, pp. 1-19, 2019. [http://dx.doi.org/10.1155/2019/4546863]

[41] "Fused Deposition Modelling of Natural Fibre/Polylactic Acid Composites", *J. Compos. Sci.,* vol. 1, no. 2, p. 8, 2017.

[42] V.K. Balla, J.G.D. Tadimeti, K.H. Kate, and J. Satyavolu, "3D printing of modified soybean hull fiber/polymer composites", *Mater. Chem. Phys.,* vol. 254, p. 123452, 2020. [http://dx.doi.org/10.1016/j.matchemphys.2020.123452]

[43] A.W. Fonseca Coelho, R.M. da Silva Moreira Thiré, and A.C. Araujo, "Manufacturing of gypsum–sisal fiber composites using binder jetting", *Addit. Manuf.,* vol. 29, p. 100789, 2019. [http://dx.doi.org/10.1016/j.addma.2019.100789]

[44] Y. Jiang, A.L. Yarin, and Y. Pan, "Printable highly transparent natural fiber composites", *Mater. Lett.,* vol. 277, p. 128290, 2020. [http://dx.doi.org/10.1016/j.matlet.2020.128290]

[45] A. Le Duigou, A. Barbé, E. Guillou, and M. Castro, "3D printing of continuous flax fibre reinforced biocomposites for structural applications", *Mater. Des.,* vol. 180, p. 107884, 2019. [http://dx.doi.org/10.1016/j.matdes.2019.107884]

[46] A.S. Mangat, S. Singh, M. Gupta, and R. Sharma, "Experimental investigations on natural fiber embedded additive manufacturing-based biodegradable structures for biomedical applications", *Rapid Prototyping J.,* vol. 24, no. 7, pp. 1221-1234, 2018. [http://dx.doi.org/10.1108/RPJ-08-2017-0162]

[47] R. Matsuzaki, M. Ueda, M. Namiki, T.K. Jeong, H. Asahara, K. Horiguchi, T. Nakamura, A. Todoroki, and Y. Hirano, "Three-dimensional printing of continuous-fiber composites by in-nozzle impregnation", *Sci. Rep.,* vol. 6, p. 23058, 2016. [http://dx.doi.org/10.1038/srep23058] [PMID: 26965201]

[48] L. Sang, S. Han, Z. Li, X. Yang, and W. Hou, "Development of short basalt fiber reinforced polylactide composites and their feasible evaluation for 3D printing applications", *Compos., Part B*

Eng., vol. 164, pp. 629-639, 2019.
[http://dx.doi.org/10.1016/j.compositesb.2019.01.085]

[49] J. Suteja, H. Firmanto, A. Soesanti, and C. Christian, "Properties investigation of 3D printed continuous pineapple leaf fiber-reinforced PLA composite", *J. Thermoplast. Compos. Mater,* 2020.
[http://dx.doi.org/10.1177/0892705720945371]

[50] H. Zhang, D. Liu, T. Huang, Q. Hu, and H. Lammer, "Three-dimensional printing of continuous flax fiber-reinforced thermoplastic composites by five-axis machine", *Materials (Basel),* vol. 13, no. 7, p. E1678, 2020.
[http://dx.doi.org/10.3390/ma13071678] [PMID: 32260222]

CHAPTER 9

Investigation of Ionic Transport in Ionic Liquid $BMIMBF_4$ Incorporated in ZnBDC-Metal-Organic Framework Based PVdF-HFP Composite Membranes

Rituraj Dutta[1,*] and **Ashok Kumar**[1]

[1] *Department of Physics, Tezpur University, Tezpur, Napaam-784028, Assam, India*

Abstract: ZnBDC-metal-organic framework (ZnBDC-MOF) has been synthesized by the solvothermal method by incorporating ionic liquid (IL) 1-butyl-3-methylimidazolium tetrafluoroborate ($BMIMBF_4$) at different wt%. The IL incorporated ZnBDC-MOF nanocomposites have been dispersed as composite nanofillers in poly (vinylidene fluoride-co-hexafluoropropylene) (PVdF-HFP) to synthesize composite polymeric membranes. XPS has investigated differences in binding energies of Cu spin-orbit peaks upon the incorporation of IL. The structural features of the Zn metal nodes of ZnBDC-MOF based composite membranes have been analysed upon IL incorporation through scanning XANES and EXAFS. Temperature-dependent ionic conductivity and electrochemical window of the IL incorporated ZnBDC-MOF based PVdF-HFP composite electrolytes have also been investigated in this chapter to study the dynamics of ion transport in the composite polymeric membranes.

Keywords: Electrochemical window, Ionic conductivity, Ionic liquid, Metal-organic framework, Polymeric membranes.

INTRODUCTION

Metal-Organic Frameworks (MOFs) are hybrid, microporous, and novel coordinating compounds consisting of inorganic unsaturated metal nodes, clusters associated with organic ligands or linker molecules [1 - 3]. In the past few decades, tremendous research has been conducted on the synthesis and applications of MOFs-based composite materials due to their tunable framework topology, controlled pore features, and promising storage capacity [4 - 7]. Owing to their excellent coordination geometry as well as porous topology, MOFs have

* **Corresponding author Rituraj Dutta:** Department of Physics, Tezpur University, Tezpur, Napaam-784028, Assam, India; E-mail: riturajdutta2009@gmail.com

C. Samson Jerold Samuel, M. Suresh & S. Gnanasekaran (Eds.)

widely been used nanomaterials for different gas storage, sensing, catalytic and sequestration applications [8 - 10]. MOFs are considered suitable host materials for encapsulating guest molecules within their pores and controlling their phase dynamics by tunable guest-host interaction. Ionic Liquids (ILs) are organic-inorganic molten salts with asymmetric cations and symmetric anions that may possess a variety of structural geometry [11, 12]. ILs can be embedded in the pores of MOFs to control the phase dynamics of ions of ILs through interaction with the metal centers and organic ligands of host frameworks. Several researchers have reported the incorporation of IL in metal-organic frameworks (MOFs) for their application as ionic conducting composite electrolytes in rechargeable batteries. Li-Hong Chen *et al.* [13] investigated the incorporation of IL EMIMCl in the UiO-67 framework and obtained ionic conductivity of 1.67×10^{-3} S cm^{-1} at 200 °C. Kazuyuki Fujie *et al.* [14] witnessed improved ionic conductivity at low temperature upon incorporating IL EMI-TFSA in the ZIF-8 framework. Qiuxia Xu *et al.* [15] reported the incorporation of IL in Cr based MOF MIL-101 and obtained ionic conductivity of 6.21×10^{-3} $S{\cdot}cm^{-1}$ at 150°C. EXAFS and XANES are significant techniques to analyze the coordination structure of the inorganic unsaturated metal nodes or clusters of MOFs and their guest-host interaction with the incorporated molecules. C. Prestipino *et al.* [16] investigated the structural topology of the MOF HKUST-1 at Cu K-edge, and L. Sciortino *et al.* [17] investigated the coordination geometry of FeBTC-MOF at Fe K-edge using XANES and EXAFS techniques. P K-edge XANES spectra of MOF functionalized with phosphine and phosphine oxide were investigated by F. L. Morel *et al.* [18] to study the topological geometry of the framework. Using XPS analyzer, the elemental composition of the Ni-based framework was investigated by W. Lu *et al.* [19]. Grafting of IL in Cr-based framework was reported by Y. Sun *et al.* [20] to improve the capability of fixation of CO_2.

Polymer electrolytes are ionically conducting solvent-free systems formed by dissolving low lattice energy salts in a polar polymer matrix [21]. Polymer electrolytes are key components of electrochromic devices as they possess several advantages such as liquid-like degrees of freedom as well as wide electrochemical window [22]. The electrochemical properties of polymer electrolytes can be influenced by inducing size-dependent filler particles. The crystallization rate of polymer segments can be suppressed by adding a high concentration of nano-sized fillers [23 - 25]. Several researchers have reported the incorporation of MOFs as efficient nanofillers in polymer electrolytes to enhance the electrochemical performance of the composite membranes for their applications in energy storage devices. The development of PEO-based composite electrolytes was reported by C. Gerbaldi *et al.* [29]. That study revealed improved temperature-dependent ionic conductivity with interfacial and mechanical stability upon controlled incorporation of nano-sized Al-BTC MOF. The electrochemical cell with the synthesized polymeric membrane delivered 40 mA h g^{-1} specific capacity at 5 C. The electrochemical cell exhibited excellent electrochemical performance with 100 mAh g^{-1} specific capacitance at 2 C after completing 100 cycles. Fabrication of Li-ion conducting MOF-based electrolytes has been reported by Nan Chen *et al.* [26] that showed the capability of dendrite-free Li deposition at high temperatures. The facile transport of Li^+ ions through the electrolytes was controlled by the electrostatic interaction of $TFSI^-$ anions with the metal nodes of MOF. N. Angulakshmi *et al.* [27] synthesized the PEO-based polymeric membranes by incorporating Mg-BTC MOF and LiTFSI that revealed enhanced ionic conductivity with discharge capacitance of 110 mAh g^{-1} at a current rate of 1 C at 70 °C. H. Huo *et al.* [28] investigated the incorporation of -NH_2 functionalized cationic MOF in PEO-based polymeric membranes to immobilize anions to enrich the Li^+ ion conduction through the free volume channels of polymer segments. The composite electrolytes exhibited a wide electrochemical window up to 4.97 V with a Li^+ transference number of 0.72. The fabricated electrochromic cell with the $LiFePO_4$ cathode materials revealed 85.4% capacity retention at 1 C after 300 cycles. R. Kumar *et al.* [29] investigated the effect of incorporation of Cu-BDC MOF in LiTFSI and PEO based polymeric membranes that revealed temperature-dependent enhancement in ionic conductivity. The electrochemical cell based on the MOF embedded electrolyte membranes delivered 98% coulombic efficiency with discharge capacitance of 120 mAh g^{-1} at 1C.

Ion conduction in polymer electrolytes is restricted by the mobility of cations as well as anions of IL as neutral ion pairs are formed by their electrostatic interaction. The mobile anions can also generate interfacial passivation (SEI) layer caused by irreversible oxidation that decreases the anodic stability. Controlled incorporation of ILs in the pores of MOFs can enhance the cationic transport through the MOF channels as the inorganic metal clusters confine the anions of IL by electrostatic interaction. The cations of IL residing at the open pores of MOFs near the linker molecules are free to migrate to the adjacent polymer segments through the channels of MOFs. Both intrachain and interchain ion hopping can occur in polymeric segments. The free cations can perform unlimited forward and backward hoping to intensify the ionic conduction in the composite electrolyte membranes.

In this chapter, IL $BMIMBF_4$ has been incorporated in ZnBDC-MOF at different wt% of IL. The IL incorporated ZnBDC-MOF composites are dispersed in PVdF-HFP polymer electrolytes to synthesize the composite polymeric membranes. The effect of IL incorporation on coordination geometry and bond elongation of ZnBDC-MOF based electrolyte membranes has been analyzed by scanning XANES and EXAFS. The binding energy differences of the Zn^{2+} species before and after IL incorporation in ZnBDC-MOF based PVdF-HFP membranes have been investigated by XPS analyzer. The AC ionic conductivity and anodic stability of electrochemical windows have also been studied to carry out the ion transport mechanism of IL $BMIMBF_4$ incorporated ZnBDC-MOF based PVdF-HFP composite polymeric membranes.

MATERIALS AND METHODS

For synthesizing the composite polymeric membranes, IL $BMIMBF_4$ was purchased from Merck, and PVdF-HFP was procured from Sigma Aldrich. The PVdF-HFP copolymer consists of vinylidene fluoride (VdF) and hexafluoropropylene (HFP) units. The VdF units provide mechanical strength and chemical stability, while the HFP units reduce the rate of crystallinity for facile ion conduction through the polymer segments. For the synthesis of ZnBDC-MOF, Zinc nitrate hexahydrate ($Zn(NO_3)_2.6H_2O$) was used as a metal precursor, and 1,4-benzene dicarboxylic acid (H_2BDC) was used as a linker and purchased from Avantor Performance Materials and Sigma Aldrich, respectively. The solvents such as N, N-dimethylformamide (DMF), Triethylamine, Ethanol, Methanol, and Acetonitrile were purchased from Merck.

Solvothermal synthesis route has been followed to prepare ZnBDC-MOF [30]. 4.5 mmol of Zinc nitrate hexahydrate and 1.5 mmol of 1,4-benzene dicarboxylic acid (H_2BDC) were dissolved consequently in 30 ml DMF. After room temperature stirring for 30 minutes, 1 ml Triethylamine was injected into the reactor. The solution was stirred for 24 h at room temperature then filtered and washed with 100 ml DMF to remove the unreacted zinc nitrate part from the solution. Finally, the solution was washed with methanol and then dried under vacuum at 150° C. After drying for 12 h, the white-coloured yield was obtained. IL $BMIMBF_4$ has been incorporated into ZnBDC-MOF at 30 wt%, 40 wt%, and 50 wt% by direct mixing method using mortar and pestle. For homogeneous incorporation of IL into the pores of ZnBDC-MOF, the nanocomposites were heated at 100°C under vacuum for 12 h. By solution casting method IL incorporated ZnBDC-MOF based polymeric membranes were prepared. 1g of PVdF-HFP was dissolved in 20 ml of Acetonitrile and ultrasonicated for 2 h to get a homogeneous mixture. Then, 0.5 g of pristine ZnBDC-MOF and each IL incorporated ZnBDC-MOF nanocomposites were dispersed separately in the polymer matrix under ultrasonic treatment for 2 h. The solutions were cast on glass slides, and after drying at room temperature for 3 to 4 h, white flexible films with thickness 20-30 μm were obtained.

RESULTS AND DISCUSSION

X-ray Diffraction Analysis

The XRD patterns of pristine ZnBDC-MOF and pristine PVdF-HFP have been provided in Fig. **1(a)**, and that of the different wt% of IL incorporated ZnBDC-MOF based polymeric membranes were depicted in Fig. **1(b)**. In Fig. **1(a)**, the characteristic peak at $2\theta = 6^{\circ}$ corresponds to the (200) planes, $2\theta = 9^{\circ}$ corresponds to the (220) planes, $2\theta = 13.7^{\circ}$ corresponds to the (400) planes and $2\theta = 15.5^{\circ}$ corresponds to (420) planes of ZnBDC-MOF. The peak at $2\theta = 19.7^{\circ}$ corresponds to the (020) planes of PVdF-HFP.

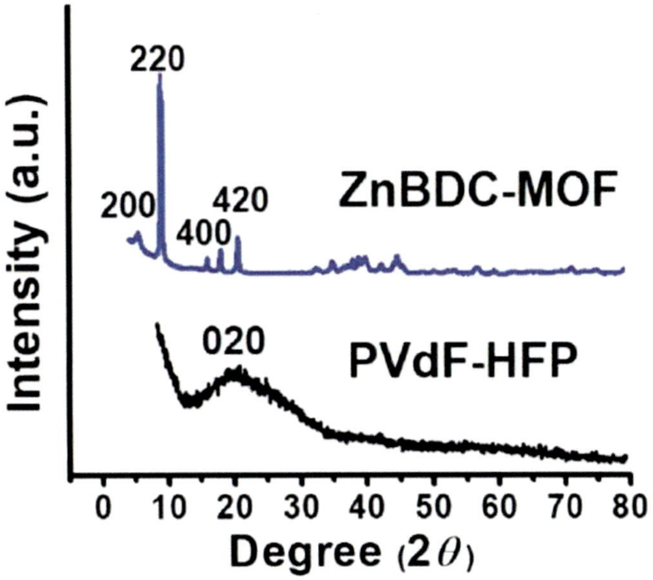

Fig. (1a). X-Ray Diffraction patterns of ZnBDC-MOF and PVdF-HFP.

The XRD patterns of ZnBDC-MOF based PVdF-HFP polymeric membranes upon incorporating $BMIMBF_4$ at different wt% are depicted in Fig. (**1b**). The intensity of the composite polymeric membranes decreases with increasing IL content as the mobile cations of IL can plasticize the nearby PVdF-HFP segments, increasing the amorphicity of the composite membranes. The % degree of crystallinity Xc has been carried out using Eq. (1) for pristine and IL $BMIMBF_4$ incorporated ZnBDC-MOF based PVdF-HFP polymeric membranes.

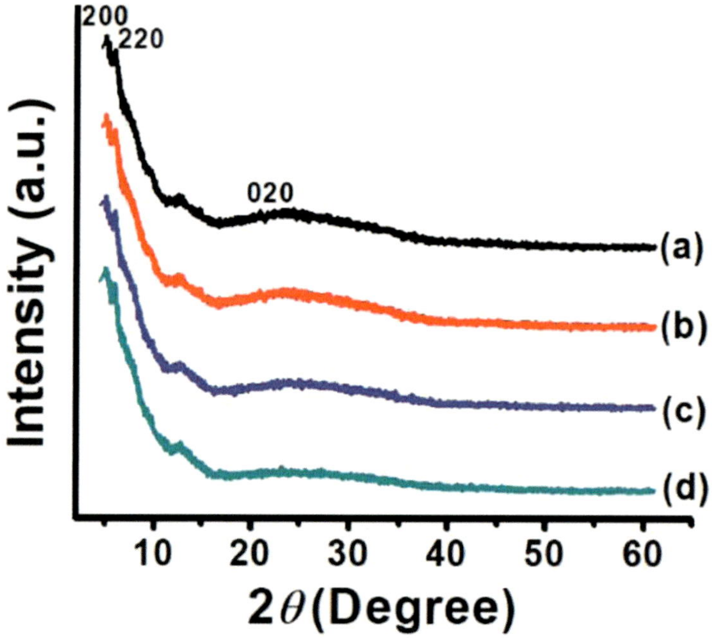

Fig. (1b). XRD patterns of ZnBDC-MOF based PVdF-HFP membranes upon different wt% of $BMIMBF_4$ incorporation, where (a) 0 wt%, (b) 30 wt%, (c) 40 wt% and (d) 50 wt%.

$$X_c = \left(\frac{A}{A_T}\right) . 100\ \% \quad (1)$$

where, A_0 is the whole area of the diffractogram, and *A* is the area of a certain crystalline peak.

The degree of crystallinity decreases from 26.4% to 15.7% at 50 wt% of IL incorporation in ZnBDC-MOF based PVdF-HFP polymeric membranes. Confinement of anions of IL in the ZnBDC framework can intensify the cation mobility that can plasticize the polymer segments in the PVdF-HFP polymeric membranes, leading to minimizing the degree of crystallinity.

X-ray Photoelectron Spectroscopy (XPS) Analysis

X-ray photoelectron Spectroscopy (XPS) is an analytical technique to analyze the chemical state, elemental composition as well as binding energy of materials. X-rays of monoenergetic Al Kα or Mg Kα are typically used XPS analyzer to penetrate the experimental material for emission of photoelectrons from the surface under vacuum treatment of $\leq 5 \times 10^{-10}$ mbar to avoid surface contamination for acquiring accurate data. The binding energy of the electron (*BE*) can be calculated using the electron's kinetic energy (*KE*) and work function (φ) of the spectrometer as follows,

$$BE = h\nu - KE - \varphi \quad (2)$$

The ESCALAB Xi⁺ XPS analyzer has been used to investigate the elemental composition and the binding energy states of different species present in the ZnBDC-MOF based PVdF-HFP composite polymeric membranes.

The XPS spectra of ZnBDC-MOF based polymer electrolytes and 50 wt% of IL $BMIMBF_4$ incorporated ZnBDC-MOF based polymer electrolytes are depicted in Fig. (**2a**). XPS data confirm the availability of Zn, C, F, and O in ZnBDC-MOF based PVdF-HFP electrolytes, while B and N are also available at the 50 wt% of IL incorporated one.

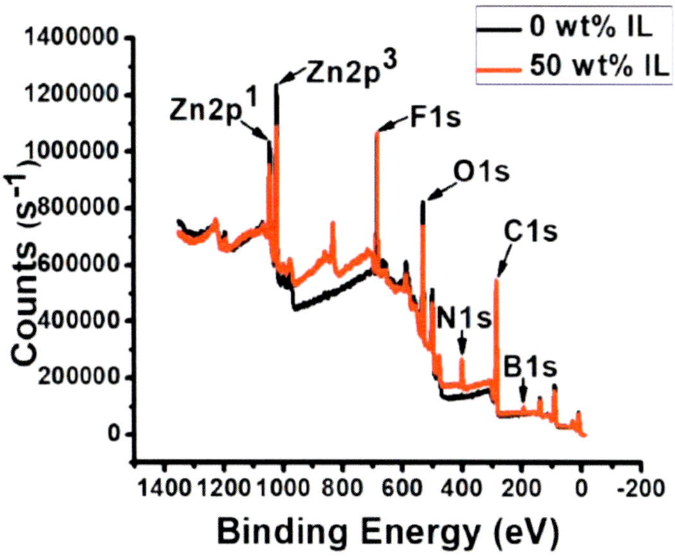

Fig. (2a). XPS spectra of pristine and 50 wt% $BMIMBF_4$ incorporated ZnBDC-MOF based PVdF-HFP polymeric membranes.

The spin-orbit peaks corresponding to Zn $2p^{1/2}$ and Zn $2p^{3/2}$ appear respectively at 1046.1 and 1026.2 eV, while C 1s peak appears at 287.4 eV, indicating the benzene-1,4 dicarboxylic acid linker. The characteristic peak of O 1s appears at 531.6 eV, indicating an unsaturated Zn-O metal node. The peak of F 1s appears at 685.3 eV, indicating PVdF-HFP. With IL loading, the characteristic spin-orbit peaks of Zn species have shifted towards higher binding energy positions, as depicted in Fig. (**2b**). Differences in binding energies of 0.8 eV and 0.67 eV have been observed for the Zn $2p^{1/2}$ and Zn $2p^{3/2}$ XPS peaks as they shift to 1046.9 and 1026.87 eV, respectively, upon 50 wt% IL incorporation. This is attributed to the interaction of Zn metal nodes with BF_4^- anions of IL. As a result, the anions of IL $BMIMBF_4$ tend to move towards the Zn nucleus that in turn increases the corresponding binding energies.

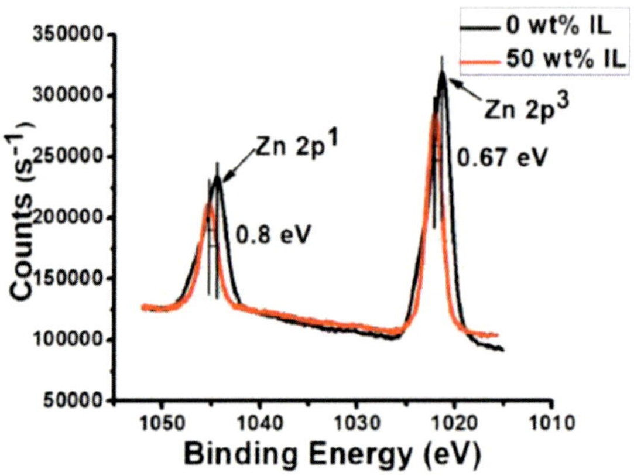

Fig. (2b). XPS spectra for Zn 2p peaks of pristine and 50 wt% IL $BMIMBF_4$ incorporated ZnBDC-MOF based PVdF-HFP polymeric membranes.

Scanning EXAFS and XANES Analysis

Scanning EXAFS and XANES techniques have been adopted to investigate the local coordination structure of the Zn K-edge of pristine and IL $BMIMBF_4$ incorporated ZnBDC-MOF based polymer electrolytes. The scanning EXAFS and XANES data were acquired by a 4.25 keV synchrotron beamline operating in transmission mode with the flux of 10^{11} photon/s. In the synchrotron beamline, a 2.5 GeV bending magnet was used as the photon source for the beam size ($H \times V$) of 1 mm × 0.2 mm with an energy resolution ($E / \Delta E$) of 10^4. In transmission mode scanning EXAFS measurement, the absorption coefficient $\mu(E)$ of any material is expressed as,

$$I = I_0 \exp [-\mu(E)x] \tag{3}$$

where, I is the photon beam that is transmitted from the surface, I_0 is the incident beam that travels a distance x. The scanning EXAFS spectra were recorded in the range from 9600 to 10800 eV, and scanning XANES spectra were observed from 9750 to 10000 eV. For energy and angle calibrations, the absorption spectrum of zinc foil has been recorded as reference one prior to the experimental samples.

In the EXAFS spectra in Fig. (**3a**), the K-edge peak of ZnBDC-MOF observed at 9827.1 eV can confirm the +2 oxidation of Zn species attributed to the Zn-Zn first shell of zinc foil used for calibration as reference material. With increasing wt% of IL incorporation, the absorption coefficients at Zn K-edges increase from 1.22 to 6.05. In the XANES spectra in Fig. (**3b**), the resonance peak appears just after the K edge peak, which can be attributed to the Zn-O bonding in ZnBDC-MOF.

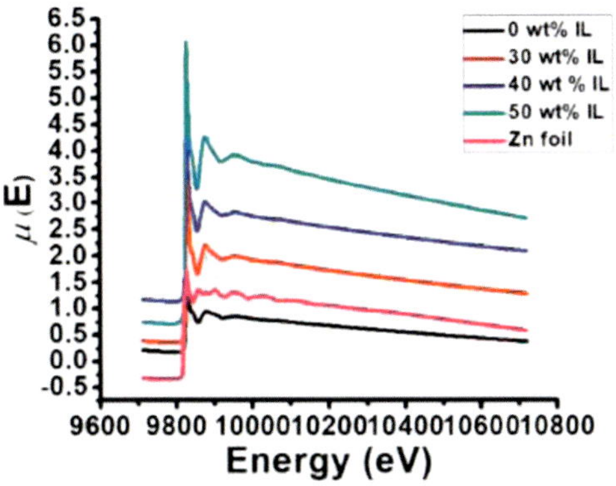

Fig. (3a). Scanning EXAFS spectra of pristine and IL incorporated ZnBDC-MOF based PVdF-HFP polymeric membranes.

Change in absorption coefficients of Zn K-edge peaks with increasing IL content in ZnBDC-MOF based polymeric membranes can be attributed to the heterogeneity in periodicity owing to the first shell degradation of Zn-O by the interaction BF_4^- anions with Zn nodes.

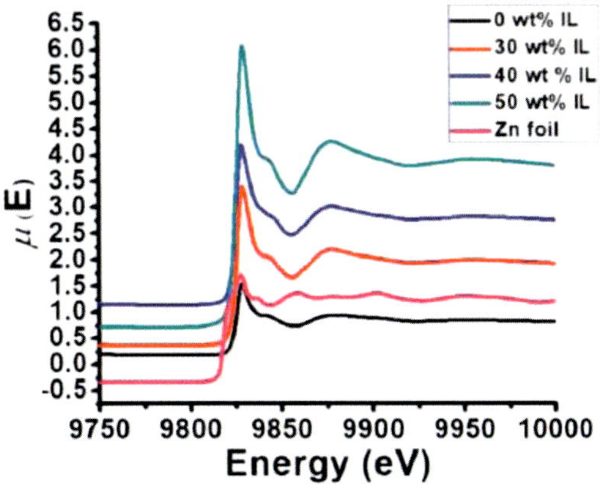

Fig. (3b). Scanning XANES spectra of pristine and IL incorporated ZnBDC-MOF based PVdF-HFP polymeric membranes.

Thermogravimetric Analysis

The thermogravimetric (TG) plot of pristine ZnBDC-MOF is depicted in Fig. (**4a**). TG plots for PVdF-HFP and different wt% of IL incorporated ZnBDC-MOF based PVdF-HFP composite polymer electrolyte membranes have been depicted in Fig. (**4b**).

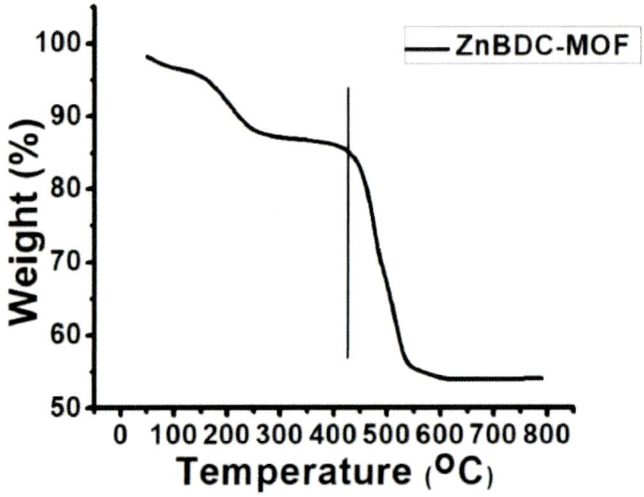

Fig. (4a). TG curve of ZnBDC-MOF.

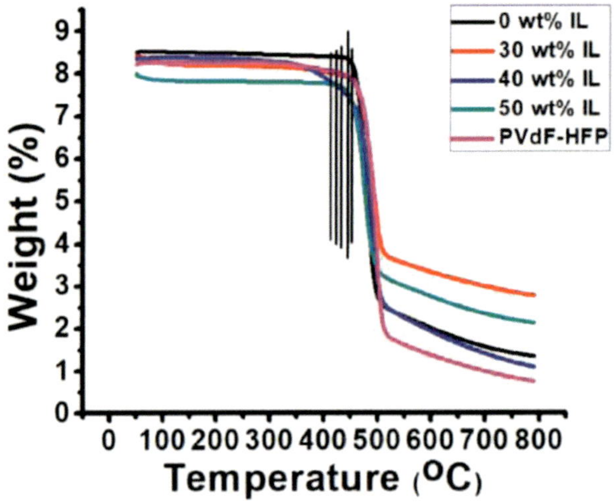

Fig. (4b). TG curves of PVdF-HFP and different wt% of IL incorporated ZnBDC-MOF based PVdF-HFP composite polymer electrolytes.

The major degradations of ZnBDC-MOF and PVdF-HFP occur at 424.6 and 451.2 °C, respectively. The degradation slope for ZnBDC-MOF merges with the slope of major degradation of PVdF-HFP, as ZnBDC-MOF has already been dispersed in the polymer electrolyte matrix to synthesize the composite polymer electrolyte membranes. With increasing wt% of IL in ZnBDC-MOF, a reduction in thermal stability is observed as the amorphicity of the composite polymer electrolyte membranes increases with the increasing incorporation of IL. The degradation temperature reduced up to 411.4 °C at 50 wt% of IL incorporation in ZnBDC-MOF based composite polymer electrolytes, as depicted in Table 1. The derivative TG plot of pristine ZnBDC-MOF is depicted in Fig. (**4c**). Derivative TG plots for PVdF-HFP and different wt% of IL incorporated ZnBDC-MOF based PVdF-HFP composite polymer electrolyte membranes are depicted in Fig. (**4d**).

Table 1. Degradation temperatures of ZnBDC-MOF, PVdF-HFP, and different wt% of IL incorporated ZnBDC-MOF based PVdF-HFP composite polymer electrolytes.

Sample	Degradation Temperature (°C)
ZnBDC-MOF	424.6
PVdF-HFP	451.2
0 wt% IL	443.6
30 wt% IL	430.4
40 wt% IL	419.1
50 wt% IL	411.4

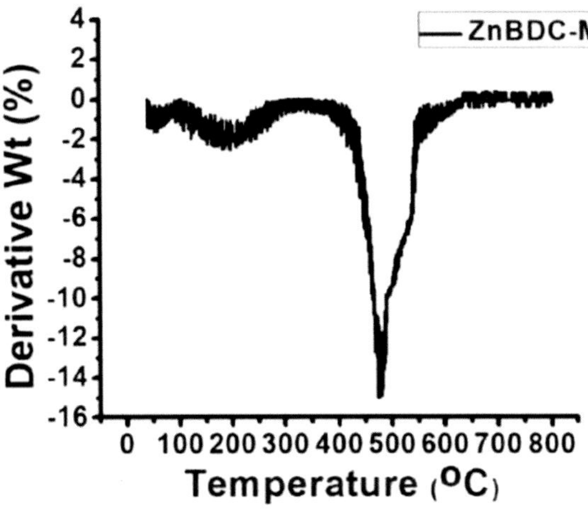

Fig. (4c). Derivative TG curves of ZnBDC-MOF.

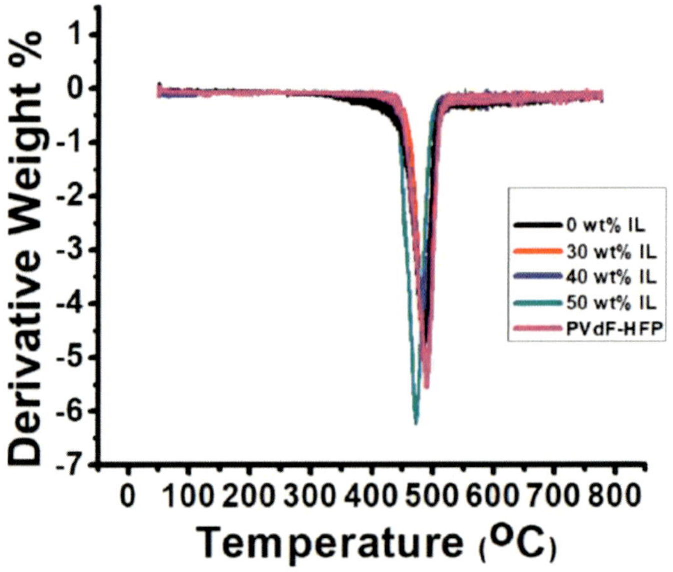

Fig. (4d). Derivative TG curves of PVdF-HFP and different wt% of IL incorporated ZnBDC-MOF based PVdF-HFP nanocomposites.

It is observed from the derivative of TG plots pristine ZnBDC-MOF exhibits the onset and most rapid temperatures at 388.9 and 473.4 °C, respectively, and pristine PVdF-HFP exhibits the onset and most rapid temperatures at 447.4 and 499.2 °C, respectively.

The onset and most rapid temperatures are observed to reduce to 426.3 and 481.4 °C, respectively, upon incorporating 50 wt% of IL in ZnBDC-MOF based PVdF-HFP membranes due to increased amorphicity in the composite polymer electrolytes, and the values are depicted in Table **2**.

Table 2. Onset and most rapid temperatures of different wt% of IL incorporated ZnBDC-MOF based PVdF-HFP composite polymer electrolytes.

Sample	Onset Temperature (°C)	Rapidest Temperature (°C)
0 wt% IL	441.6	496.6
30 wt% IL	436.7	491.4
40 wt% IL	431.4	486.7
50 wt% IL	426.3	481.4

Ionic Conductivity Analysis

The impedance analysis of the ZnBDC-MOF based composite polymeric membranes have been monitored from 42 Hz to 5 MHz by dielectric LCR Hitester made of Hioki 3532-50 at different temperatures up to 380 K. The magnitude of the impedance ($|Z| = V_{max}/I_{max}$) with phase angle (θ) can be illustrated as functions of the corresponding angular frequency.

The room temperature Nyquist plots of different wt% of IL $BMIMBF_4$ incorporated ZnBDC-MOF based PVdF-HFP polymeric membranes are depicted in Fig. (**5a**).

From the Nyquist impedance plots, the ionic conductivities of different wt% of IL $BMIMBF_4$ incorporated ZnBDC-MOF based polymeric membranes are investigated at 300 K and obtained 1.3×10^{-3} S cm^{-1} at 50 wt% of IL. The variation of ionic conductivity at 300 K at different wt% of IL in ZnBDC-MOF based composite membranes have been depicted in Fig. (**5b**). Controlled incorporation of IL up to 50 wt% in the pores of ZnBDC-MOF can enrich the cation mobility through the free channels of ZnBDC-MOF as the BF_4^- anions are closely packed or confined by electrostatic interaction with the secondary building blocks of Zn nodes. The imidazolium cations of IL can interact with the BDC organic molecules and stay at the open pores to migrate to the adjacent PVdF-HFP polymer segments. As intrachain as well as interchain motions are possible in the vicinity of the polymeric segments, the cations can perform random forward-backward hopping to the adjacent coordination sites through the interconnected polymer chains resulting in increased ionic conductivity with growing wt% for IL incorporation in ZnBDC-MOF.

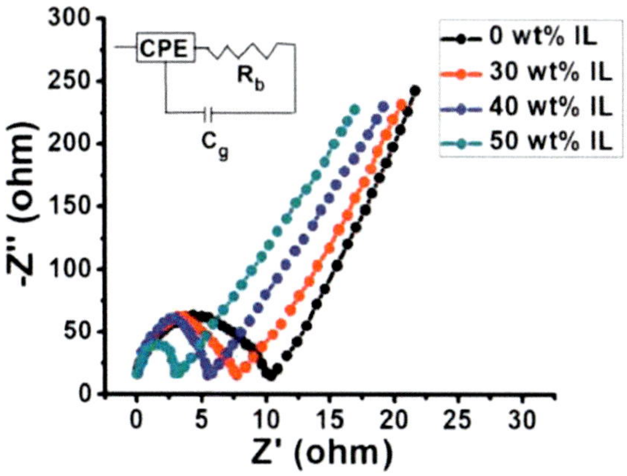

Fig. (5a). Nyquist plots of different wt% of IL $BMIMBF_4$ incorporated ZnBDC-MOF based PVdF-HFP polymeric membranes at 300K.

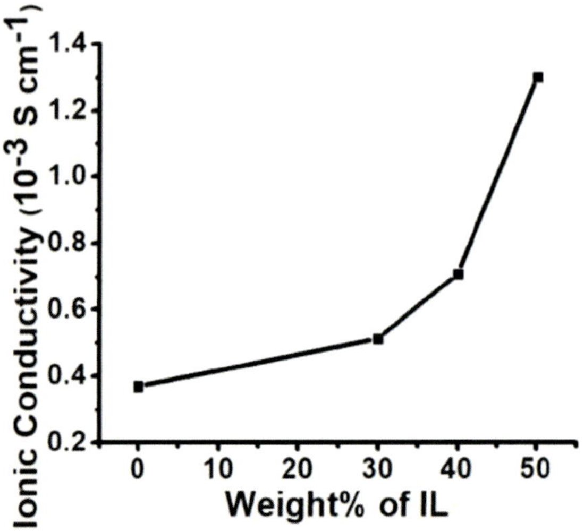

Fig. (5b). Ionic conductivity at room temperature for different wt% of IL incorporated ZnBDC-MOF based PVdF-HFP membranes.

Temperature-dependent plots of ionic conductivity (σ) for different wt% of $BMIMBF_4$ incorporated ZnBDC-MOF based polymeric membranes have been observed from 300 K to 380 K as presented in Fig. (**5c**).

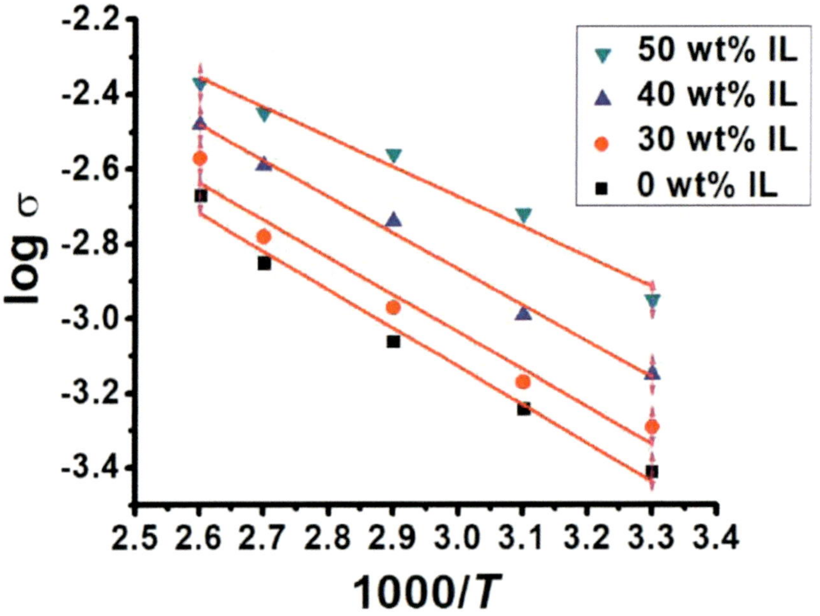

Fig. (5c). Temperature-dependent ionic conductivity plots of IL incorporated ZnBDC-MOF based PVdF-HFP membranes.

With increasing temperature, the ion mobility increases with an enhanced segmental motion of PVdF-HFP membranes, and conductivity is obtained as 3.3×10^{-3} Scm^{-1} at 380 K. Due to the expansion of polymer segments with increasing temperature and more free volume, pathways are generated for easy cationic transport throughout the electrolytes, leading to intensifying ionic conductivity [31, 32].

Electrochemical Stability Analysis

The electrochemical window profiles of different wt% of $BMIMBF_4$ incorporated ZnBDC-MOF based PVdF-HFP membranes are obtained from Linear Sweep Voltammetry measurements using Autolab galvanostat/potentiostat (Model: PGSTAT 302 N) as depicted in Fig. (**6**). The experiment was carried out at room temperature using two electrodes workstation setup by joining a stainless steel reference electrode with the counter on one side of the electrochemical cell and the stainless steel working electrode on the other side at a scan rate of 0.1 V s^{-1}.

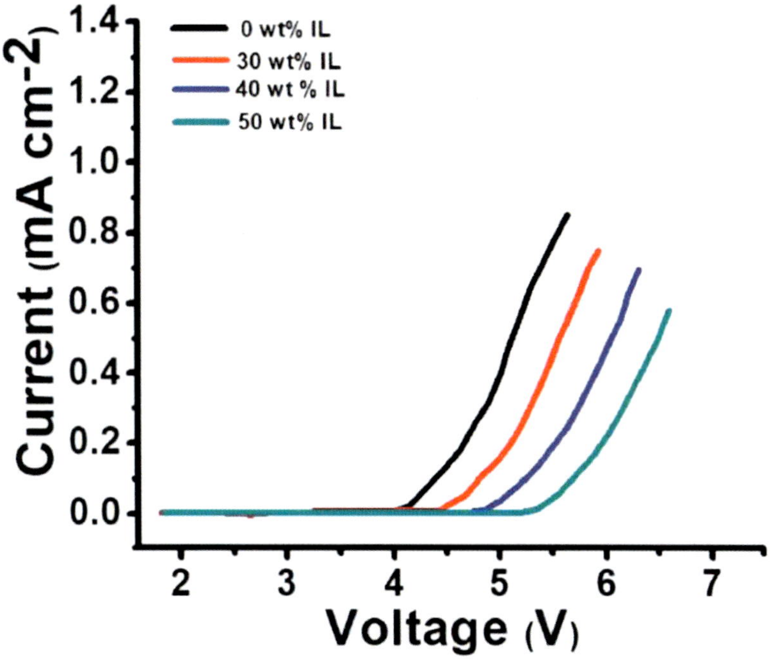

Fig. (6). Electrochemical window of different wt% of $BMIMBF_4$ incorporated ZnBDC-MOF based PVdF-HFP membranes.

For the pristine ZnBDC-MOF based electrolytes without IL content, no current is observed up to 3.7 V throughout the working electrode, and current is started to increase beyond 3.7 V. The electrochemical window widens with increasing content of IL and attains 5.2 V at 50 wt% of IL incorporation in the ZnBDC-MOF based polymeric membranes. In polymer electrolytes, irreversible anionic oxidation at anode-electrolyte interfaces can suppress the anodic stability [33]. Interaction of BF_4^- anions of IL with the ZnBDC-MOF can shield the anionic contact with the electrodes that minimize the interfacial oxidation reaction of anodes leading to increase electrochemical stability.

CONCLUDING REMARKS

In this chapter, the ion transport characteristics of ZnBDC-MOF based PVdF-HFP polymeric membranes have been studied upon $BMIMBF_4$ incorporation in ZnBDC-MOF. From the XPS spectra, Zn $2p^{1/2}$ and Zn $2p^{3/2}$ peaks are observed to shift by 0.8 eV and 0.67 eV, respectively, suggesting the movement of BF_4^- anions towards the Zn nodes at 50 wt% of $BMIMBF_4$ incorporated ZnBDC-MOF based PVdF-HFP membranes. The K-edge absorption coefficients in scanning EXAFS data are observed to increase from 1.22 to 6.05 upon increasing IL content in ZnBDC-MOF, suggesting the confinement of BF_4^- anions in the MOF pores. From thermogravimetric analysis, degradation in thermal stability is observed from 443.6 to 411.4 °C upon increasing incorporation of IL in ZnBDC-MOF based composite polymer electrolyte membranes. Reduction in thermal stability can be attributed to the increase in amorphicity of the composite polymer electrolyte membranes upon increasing wt% of incorporation of IL. Ionic conductivity of 1.3×10^{-3} S cm^{-1} is observed at 300 K upon a maximum of 50 wt% of $BMIMBF_4$ incorporation in ZnBDC-MOF based polymeric membranes. Ionic conductivity enhances with increasing temperature owing to enhanced mobility of cations throughout the polymer backbones and attains 3.3×10^{-3} S cm^{-1} at 380 K. The electrochemical window increases from 3.7 V for without IL incorporation to 5.2 V for 50 wt% of $BMIMBF_4$ incorporated ZnBDC-MOF based PVdF-HFP membranes as carried out by Linear Sweep Voltammetry measurements.

CONSENT FOR PUBLICATION

Not applicable.

CONFLICT OF INTEREST

The author declares no conflict of interest, financial or otherwise.

ACKNOWLEDGEMENTS

The authors sincerely acknowledge the financial support from the DST-INSPIRE fellowship (Grant No: DST/INSPIRE Fellowship/ 2015/ IF150994). The authors are thankful to SAIF-NEIST, Jorhat, for allowing the XPS facility. The authors also acknowledge Prof. S. N. Jha, RRCAT, Indore, for providing a synchrotron beamline for Scanning EXAFS measurements.

REFERENCES

[1] D.J. Tranchemontagne, J.R. Hunt, and O.M. Yaghi, "Room temperature synthesis of metal-organic frameworks: MOF-5, MOF-74, MOF-177, MOF-199, and IRMOF-0", *Tetrahedron,* vol. 64, no. 36, pp. 8553-8557, 2008. [http://dx.doi.org/10.1016/j.tet.2008.06.036]

[2] J.R. Li, R.J. Kuppler, and H.C. Zhou, "Selective gas adsorption and separation in metal-organic frameworks", *Chem. Soc. Rev.,* vol. 38, no. 5, pp. 1477-1504, 2009. [http://dx.doi.org/10.1039/b802426j] [PMID: 19384449]

[3] S.J. Dalgarno, P.K. Thallapally, L.J. Barbour, and J.L. Atwood, "Engineering void space in organic van der Waals crystals: calixarenes lead the way", *Chem. Soc. Rev.,* vol. 36, no. 2, pp. 236-245, 2007. [http://dx.doi.org/10.1039/b606047c] [PMID: 17264926]

[4] J.A. Bohrman, and M.A. Carreon, "Synthesis and CO2/CH4 separation performance of Bio-MOF-1 membranes", *Chem. Commun. (Camb.),* vol. 48, no. 42, pp. 5130-5132, 2012. [http://dx.doi.org/10.1039/c2cc31821k] [PMID: 22517243]

[5] Y. Chen, Z. Hu, M.G. Krishna, and J. Jiang, "Ionic liquid/metal-organic framework composite for CO_2 capture: a computational investigation"., *J. Phys. Chem. C,* vol. 115, no. 44, pp. 21736-21742, 2011. [http://dx.doi.org/10.1021/jp208361p]

[6] H. Wu, W. Zhou, and T. Yildirim, "High-capacity methane storage in metal-organic frameworks M2(dhtp): the important role of open metal sites", *J. Am. Chem. Soc.,* vol. 131, no. 13, pp. 4995-5000, 2009. [http://dx.doi.org/10.1021/ja900258t] [PMID: 19275154]

[7] A.R. Millward, and O.M. Yaghi, "Metal-organic frameworks with exceptionally high capacity for storage of carbon dioxide at room temperature", *J. Am. Chem. Soc.,* vol. 127, no. 51, pp. 17998-17999, 2005. [http://dx.doi.org/10.1021/ja0570032] [PMID: 16366539]

[8] M.P. Suh, H.J. Park, T.K. Prasad, and D.W. Lim, "Hydrogen storage in metal-organic frameworks", *Chem. Rev.,* vol. 112, no. 2, pp. 782-835, 2012. [http://dx.doi.org/10.1021/cr200274s] [PMID: 22191516]

[9] S. Couck, J.F. Denayer, G.V. Baron, T. Rémy, J. Gascon, and F. Kapteijn, An amine-functionalized MIL-53 metal-organic framework with large separation power for CO_2 and CH_4, *J. Am. Chem. Soc.,* vol. 131, no. 18, pp. 6326-6327, 2009. [http://dx.doi.org/10.1021/ja900555r] [PMID: 19374416]

[10] A.G. Wong-Foy, A.J. Matzger, and O.M. Yaghi, Exceptional H_2 saturation uptake in microporous metal-organic frameworks., *J. Am. Chem. Soc.,* vol. 128, no. 11, pp. 3494-3495, 2006. [http://dx.doi.org/10.1021/ja058213h] [PMID: 16536503]

[11] K.R. Seddon, "Ionic Liquid for Clean Technology", *J. Chem. Technol. Biotechnol.,* vol. 68, no. 4, pp. 351-356, 1997. [http://dx.doi.org/10.1002/(SICI)1097-4660(199704)68:4<351::AID-JCTB613>3.0.CO;2-4]

[12] A.R. Polu, and H.W. Rhee, "Ionic liquid doped PEO-based solid polymer electrolytes for lithium-ion polymer batteries", *Int. J. Hydrogen Energy,* vol. 42, no. 10, pp. 7212-7219, 2017. [http://dx.doi.org/10.1016/j.ijhydene.2016.04.160]

[13] L. Chen, B. Wu, H. Zhao, L. Long, and L. Zheng, "High temperature ionic conduction mediated by ionic liquid incorporated within the metal-organic framework UiO-67(Zr)", *Inorg. Chem. Commun.,* vol. 81, pp. 1-4, 2017. [http://dx.doi.org/10.1016/j.inoche.2017.04.019]

[14] K. Fujie, K. Otsubo, R. Ikeda, T. Yamada, and H. Kitagaw, "Low temperature ionic conductor: ionic liquid incorporated within a metal–organic framework. ", *Chem. Sci,* vol. 6, pp. 4306-4310, 2015.

[15] Q. Xu, X. Zhang, and S. Zeng, "Ionic Liquid Incorporated Metal Organic Framework for High Ionic Conductivity over Extended Temperature Range", *ACS Sustain. Chem.& Eng.,* vol. 7, pp. 7892-7899, 2019.

[16] C. Prestipino, L. Regli, J.G. Vitillo, F. Bonino, A. Damin, C. Lamberti, A. Zecchina, P.L. Solari, K.O. Kongshaug, and S. Bordiga, "Local Structure of Framework Cu(II) in HKUST-1 Metallorganic Framework: Spectroscopic Characterization upon Activation and Interaction with Adsorbates", *Chem. Mater.,* vol. 18, no. 5, pp. 1337-1346, 2006. [http://dx.doi.org/10.1021/cm052191g]

[17] L. Sciortino, A. Alessi, F. Messina, G. Buscarino, and F.M. Gelardi, "Structure of the FeBTC Metal–Organic Framework: A Model Based on the Local Environment Study", *J. Phys. Chem. C,* vol. 119, no. 14, pp. 7826-7830, 2015. [http://dx.doi.org/10.1021/acs.jpcc.5b01336]

[18] F.L. Morel, S. Pin, T. Huthwelker, M. Ranocchiari, and J.A. van Bokhoven, "Phosphine and phosphine oxide groups in metal-organic frameworks detected by P K-edge XAS", *Phys. Chem. Chem. Phys.,* vol. 17, no. 5, pp. 3326-3331, 2015. [http://dx.doi.org/10.1039/C4CP05151C] [PMID: 25523824]

[19] W. Lu, and X. Wu, "Ni-MOF nanosheet arrays: efficient non-noble-metal electrocatalysts for non-enzymatic monosaccharide sensing", *New J. Chem.,* vol. 42, no. 5, pp. 3180-3183, 2018. [http://dx.doi.org/10.1039/C7NJ04754A]

[20] Y. Sun, H. Huang, H. Vardhan, B. Aguila, C. Zhong, J.A. Perman, A.M. Al-Enizi, A. Nafady, and S. Ma, Facile Approach to Graft Ionic Liquid into MOF for Improving the Efficiency of CO_2 Chemical Fixation., *ACS Appl. Mater. Interfaces,* vol. 10, no. 32, pp. 27124-27130, 2018. [http://dx.doi.org/10.1021/acsami.8b08914] [PMID: 30016060]

[21] P.V. Wright, "Electrical Conductivity in Ionic Complexes of Poly (Ethylene Oxide)", *Br. Polym. J.,* vol. 7, no. 5, pp. 319-327, 1975. [http://dx.doi.org/10.1002/pi.4980070505]

[22] A. Hofmann, M. Schulz, and T. Hanemann, "Gel electrolytes based on ionic liquids for advanced lithium polymer batteries", *Electrochim. Acta,* vol. 89, pp. 823-831, 2013. [http://dx.doi.org/10.1016/j.electacta.2012.10.144]

[23] F. Croce, G.B. Appetecchi, L. Persi, and B. Scrosati, "Nanocomposite polymer electrolytes for lithium batteries", *Nature,* vol. 394, no. 6692, pp. 456-458, 1998. [http://dx.doi.org/10.1038/28818]

[24] W. Wieczorek, Z. Florjanczyk, and J.R. Stevens, "Composite polyether based solid electrolytes", *Electrochim. Acta,* vol. 40, no. 13-14, pp. 2251-2258, 1995. [http://dx.doi.org/10.1016/0013-4686(95)00172-B]

[25] C. Gerbaldi, J.R. Nair, A.M.A. Kulandainathan, R.S. Kumar, C. Ferrara, P. Mustarelli, and A.M. Stephan, "Innovative high performing metal organic framework (MOF)-laden nanocomposite polymer electrolytes for all-solid-state lithium batteries", *J. Mater. Chem. A Mater. Energy Sustain.,* vol. 2, no. 26, pp. 9948-9954, 2014. [http://dx.doi.org/10.1039/C4TA01856G]

[26] N. Chen, Y. Li, Y. Dai, W. Qu, Y. Xing, Y. Ye, Z. Wen, C. Guo, F. Wu, and R. Chen, A Li^+ conductive metal organic framework electrolyte boosts the high-temperature performance of dendrite-free lithium batteries., *J. Mater. Chem. A Mater. Energy Sustain.,* vol. 7, no. 16, pp. 9530-9536, 2019. [http://dx.doi.org/10.1039/C8TA12539B]

[27] N. Angulakshmi, R.S. Kumar, M.A. Kulandainathan, and A.M. Stephan, "Composite Polymer Electrolytes Encompassing Metal Organic FrameWorks: A New Strategy for All-Solid-State Lithium

Batteries", *J. Phys. Chem. C,* vol. 118, no. 42, pp. 24240-24247, 2014. [http://dx.doi.org/10.1021/jp506464v]

[28] H. Huo, B. Wu, T. Zhang, X. Zheng, L. Ge, T. Xu, X. Guo, and X. Sun, "Anion-immobilized polymer electrolyte achieved by cationic metal-organic framework filler for dendrite-free solid-state batteries", *Energy Storage Mater.,* vol. 18, pp. 59-67, 2019. [http://dx.doi.org/10.1016/j.ensm.2019.01.007]

[29] R.S. Kumar, M. Rajab, M.A. Kulandainathana, and A.M. Stephan, "Metal organic framework-laden composite polymer electrolytes for efficient and durable all solid-state-lithium batteries", *RSC Advances,* vol. 4, no. 50, pp. 26171-26176, 2014. [http://dx.doi.org/10.1039/C4RA03147D]

[30] Y. Changfu, L. Jie, P. Han, Y. Lai, Z. Zhang, and J. Liu, "Enhanced electrochemical performance of poly (ethylene oxide) based composite polymer electrolyte by incorporation of nano-sized metal-organic framework", *J. Power Sources,* vol. 240, pp. 653-658, 2013. [http://dx.doi.org/10.1016/j.jpowsour.2013.05.030]

[31] Z.D. Liu, Y. Feng, and W.L. Li, "High dielectric constant and low loss of polymeric dielectric composites filled by carbon nanotubes adhering BaTiO3 hybrid particles", *RSC Advances,* vol. 5, no. 37, pp. 29017-29021, 2015. [http://dx.doi.org/10.1039/C5RA00639B]

[32] S. Sinha, S.K. Chatterjee, J. Ghosh, and A.K. Meikap, "Dielectric relaxation and ac conductivity behaviour of polyvinyl alcohol-HgSe quantum dot hybrid films", *J. Phys. D Appl. Phys.,* vol. 47, no. 27, pp. 275301-275313, 2014. [http://dx.doi.org/10.1088/0022-3727/47/27/275301]

[33] C.H. Park, D.W. Kim, J. Prakash, and Y.K. Sun, "Electrochemical stability and conductivity enhancement of composite polymer electrolytes", *Solid State Ion.,* vol. 159, no. 1-2, pp. 111-119, 2003. [http://dx.doi.org/10.1016/S0167-2738(03)00025-0]

CHAPTER 10

Manufacturing Techniques and Applications of Multifunctional Metal Matrix Composites

S. Ganeshkumar[1,*] and **S. Venkatesh**[1]

[1] *Sri Eshwar College of Engineering, Coimbatore, Tamil Nadu, India*

[2] *Post-Doctoral Fellow, National Institute of Technology, Trichy, Tamil Nadu, India*

Abstract: In the last few years, the research on composite materials has remarkably been increased. The evolution and specialization of composite materials have widened into metal matrix composites, carbon-reinforced composites, *etc*. In addition to this, the machining and manufacturing process of composites have evolved tremendously. The processing techniques of the composite materials are classified into two types: liquid and solid type processing techniques. These composite materials achieve the weight reduction of the material without losing the yield stress. As a result, these composite materials are widely used in marine, aerospace, and automobile applications. To increase the fuel efficiency of automobiles, the weight of the vehicles is reduced using these composite materials. Composite materials are the combination of two or more materials to acquire the multifunctional metal matrix composites. Unlike alloys of metals, composite materials are visible, easily detachable, and easy to manufacture. Hence, these composite materials and their manufacturing techniques are emerging fields in the engineering industry. Strength, absorptivity, degrading property, impact, and wear resistance properties are improved in these composite materials to enhance the desired properties in conventional metals and alloys. This article exhibits various methods of the manufacturing process of composite materials and applications of multifunctional metal matrix composites.

Keywords: Casting, Metal Matrix composites, Molding process, Powder metallurgy, Thermo functional materials.

INTRODUCTION

When two materials are combined, it gives multiple properties to the composite material matrix, such as lightweight, electrical conductivity, good machining properties, stiffness, strength, *etc*. Glass fire composites were introduced by Owens Corning, leading to the evolution of carbon-reinforced fibers. The unsaturated polyester resins evolved in this era to enhance the engineering proper-

* **Corresponding author S Ganeshkumar:** Sri Eshwar College of Engineering, Tamil Nadu India; E-mail: ganeshkumar.s@sece.ac.in

C. Samson Jerold Samuel, M. Suresh & S. Gnanasekaran (Eds.)

ties of composite materials. Some common types of composite materials are ceramic matrix composite, metal matrix composites, reinforced concrete, cement-bonded wood, glass fiber, carbon fiber reinforced polymer, *etc*.

Weight reduction is the main advantage of composite materials. It is used in aerospace applications, automobile industries, marine applications, consumer products, *etc*. Composite materials are being used in equipment such as bicycles frames, where the weight is reduced, and strength is increased.

Carbon fiber reinforced plastics are widely used in the engineering industry to reduce weight, inherent good mechanical properties, and corrosion resistance. These composite materials are also used to transplant artificial eyes, ears, and organs in biomedical applications. Epoxy, PMMA, Titanium, and PTFE-based composites are widely used in transplants. The titanium composites are used to transplant bones, which leads to a good bond with the bones and muscles of the human body. In the automotive industry, weight is being reduced in carbon-reinforced polymers, which increases automobiles' efficiency to reduce fuel consumption. To perform various functions apart from the primary functions, multifunctional composites are used. In multifunctional composite manufacturing techniques, the excess material is eliminated to reduce the weight further by using topology and shape optimization techniques. In recent years, the engineering industry has evolved the need for multifunctional capabilities composite materials and manufacturing systems.

Thematic functional materials, Multifunctional Composites (MFC) and multifunctional manufacturing systems are evolved to perform sensing and actuation, improve mechanical strength and electrical conductivity, and reduce drag, respectively [1]. The complex structures of Multi-Functional Composite Materials (MFCM) challenge the manufacturing process. The fabrication process plays an essential role in mechanical properties by preserving initial structures and adding additional structures. The integration of functional devices exhibits the inherent mechanical behaviour within the structural materials. Embedded devices are manufactured using energy cells based on multifunctional composite material techniques. Xiao *et al.* [2] exhibited the fabrication method of energy cells using laminated composites. The thin film of lithium cells is used to store the energy in the energy cells. Crack propagation, slipping, and frictional forces are demonstrated in the experimentation. Composite materials originated in Egypt. Bricks are made with a combination of straw, which improves strength. Later, Mongolian armies made composite lightweight bows. These reduced lightweight bows travelled to longer heights based on biomaterials. The concept of composite materials is applied in the construction of buildings to inherit good mechanical properties, such as a mixture of cement and rods, leading to an increase in the tensile and compressive strength of the material. It is categorized generally by matrix and reinforcement. For example, plywood composite and wooden particles are reinforced by the binder. The binding particles are said to be the matrix. These matrices are classified as polymer, metal, and ceramic matrices. The strength, stiffness and density of the matrix are determined by reinforcement. These reinforcements can be particulates, fabrics, and whiskers.

PROCESSING OF METAL MATRIX COMPOSITES

Copper, Iron, magnesium, metals are used for medium temperatures, very high-temperature cobalt and cobalt-nickel mixture applications. Particulate reinforcement, whisker reinforcement, continuous fiber reinforcement techniques are generally used in metal matrix composites. Manufacturing techniques are classified into stir casting and infiltration. The classifications are shown in Fig. (**1**). Among the liquid and solid types of manufacturing of composite materials, solid type powder metallurgy is versatile for manufacturing and integrates good mechanical properties. Usually, there is a combination of two or more materials to improvise the material properties to meet the engineering requirements' compliance.

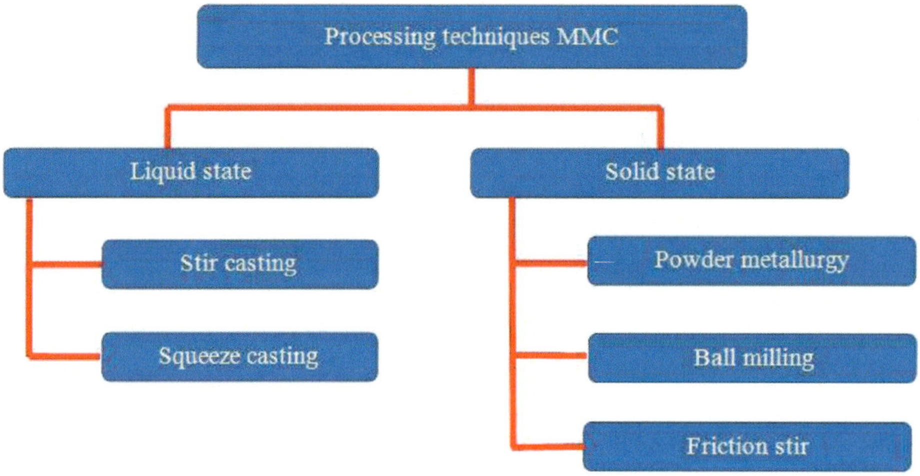

Fig. (1). Processing techniques of MMC.

The reinforcement is synthesized externally, and sintering of the powder particles of different materials gives good bond strength to the structures. Often the reinforcement is synthesized by chemical reactions of bonding agents, matrices, *etc*. Additive manufacturing is another technique for manufacturing metal matric, carbon fiber, and polymer composites. Similar to the sintering process, powder bed fusion technique, direct fusion deposition, binder jet 3D printing are techniques of additive manufacturing utilized to fabricate metal matrix and polymer matrix compounds. The composite materials are utilized in caesium adsorption. Composite materials absorb radioactive materials and emissions. Sodium carboxymethyl cellulose (CMC) is water-soluble cellulose utilized in this experimentation to adsorb radioactive emissions. Pollution due to plastic materials leads to the development of biodegradable materials [3]. Two or more biodegradable materials are utilized to reinforce the materials and increase the mechanical properties. Bio-derived polymers with improved property profiles have been developed in the engineering industry. Cellulose Nanocrystals are obtained from biomass, which reduces machining cost and an increasing rate of biodegradability.

The properties influencing the mechanical properties of the composites include the method of deriving, nature of the processing, and the nature of additive materials. Cellulose Nanocrystal fillers have several potential benefits and application areas. Carbon nanocrystals are used in water purification, also having biomedical and electronics applications. Compression moulding, pultrusion technique, injection moulding, hand lay-up, vacuum bagging, *etc.*, are common methods of manufacturing composite materials. Gholampour [4] depicted the fabrication of composite materials using various biodegradable materials. The methods of the composite manufacturing process are exhibited in Table **1**.

Table 1. Manufacturing methods of composite materials.

S.No	Composite	Manufacturing Methods
1	Woven jute fiber/epoxy composite	Compression molding
2	Unidirectional flax fiber/polyester composite	Infusion molding
3	Unidirectional flax fiber/epoxy composite	Hand lay-up
4	Unidirectional flax fiber/polyester composite	Vacuum-assisted resin infusion
5	Random-oriented oil palm fiber/polyester composite	Hand lay-up
6	Random-oriented sugarcane fiber/polyester composite	Hand lay-up
7	Coir fiber/polyester composite	Compression molding
8	Unidirectional kenaf fiber/epoxy composite	Vacuum bagging
9	Random-oriented palm leaf fiber/polypropylene composite	Injection molding
10	Unidirectional jute fiber/phenolic composite	Pultrusion technique
11	Unidirectional jute fiber/polyester composite	Compression molding
12	Random-oriented banana fiber/phenolic composite	Compression molding

Manufacturing Methods of Composite Materials - Pultrusion Process

Among all the manufacturing processes of composite materials, the pultrusion process is widely used in fiber-reinforced composites. Fiberglass roving is fed to the forming and curing dies across the resin impregnator. Often resin-soaked fiber is used to improvise the efficiency of the composites. Pulling mechanisms assists in ejecting the fabricated composite product from the system. The schematic diagram of the manufacturing of resin-impregnated fiberglass composite material is exhibited in Fig. (**2**). The pultrusion is initiated in the reciprocating pullers.

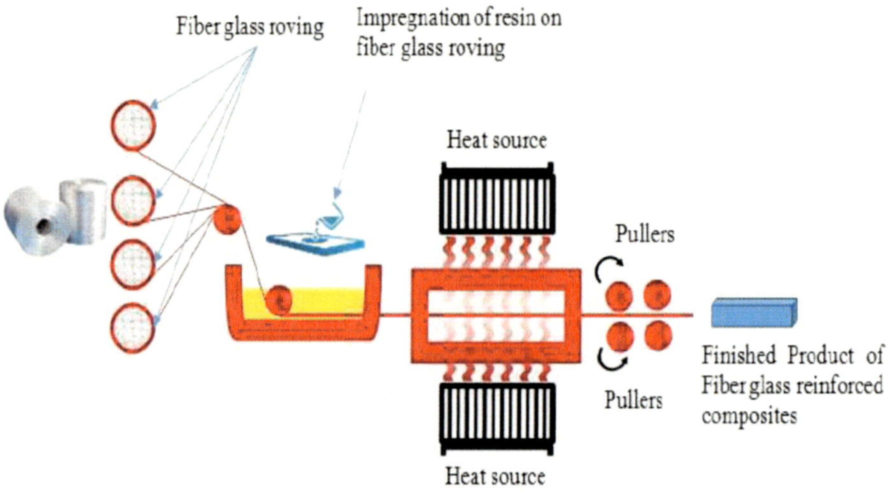

Fig. (2). Schematic Diagram of Pultrusion process of manufacturing of composites.

The surface material is added to the existing material and pulled to obtain composites. The resin-impregnated fiberglass roving is formed and cured in the die. The reciprocating mechanism continuously ejects the formed-cured composite material to a cut saw. The cut-saw cuts the composites into desired lengths of finished hardened reinforced polymer composites. In general, Fiberglass is often used as base material, and polyester resins (methyl ethyl ketone peroxide) of 75% composition are used in this pultrusion process to manufacture the fiberglass, polyester resin composite materials. Wall panels, doors, boat panels, automobile components are the major applications of fiber-reinforced resin-impregnated composites.

Compression Moulding Process

The compression moulding process is one of the manufacturing methods of composite materials. In this method, the mass production of composite material products is easily possible with high precision and accuracy. The typical compression moulding techniques employ the hot die or heated press to consolidate the fiber composite materials. The forming of composite materials is attained by a heated metal mould. The feeding composite materials are placed in open heated mould cavity. The mould is closed using a hydraulic press. Due to the pressure applied to the heated forming dies, composites reshape their size according to the geometry of forming dies. The cycle time ranges 1-5 minutes depending upon the thickness of the products. The compression moulding process is exhibited in Fig. (**3**).

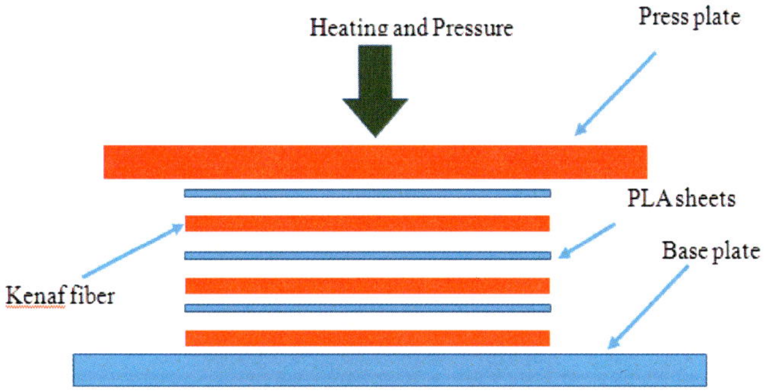

Fig. (3). Schematic Diagram of Compression moulding process of manufacturing of composites.

The pressure up to 2000 PSI is maintained in the die using a hydraulic press, leading to the generation of shear and compressive force on the composite layers. Mould coating techniques are applied to improvise the surface finish properties of compressed composite materials. Due to its simplicity, the cost of the manufacturing technique is low, and the wastage of the material is reduced. Automobile parts and window doors panels are common examples of composite materials manufactured by compression moulding [5].

Vacuum Bagging

Vacuum bagging is a very flexible and precise manufacturing process to manufacture composite materials. During the cure cycle, mechanical pressure is applied on the surface of the laminate to form a coating. This technique is called vacuum bagging. In pressuring composite laminates, several functions are involved, such as removing trapped air between the surfaces. Vacuum bagging is used to manufacture large parts such as wind turbine blades, the body of boats, *etc*. Since pressure is relatively low at 1 bar, the surface finish will be good on the tool side, and medium surface qualities will be achieved on the bag side [6]. The schematic representation of steps involved in vacuum bagging is exhibited in Fig. (**4**).

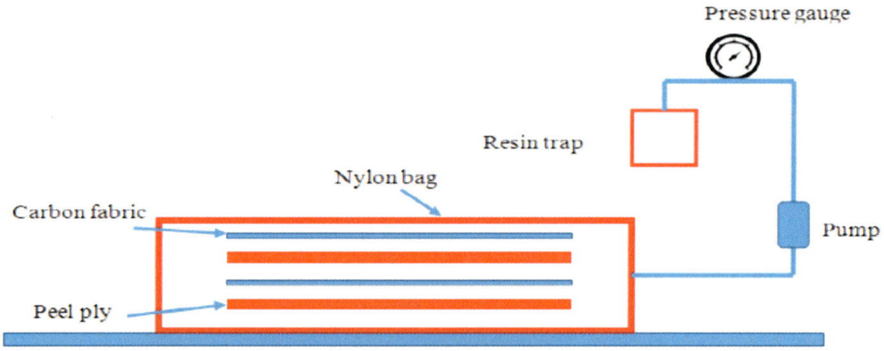

Fig. (4). Schematic Diagram of Vacuum bagging process.

The consolidation of composite layers and manufacturing process cost is very less. In the vacuum bagging method, atmospheric air is given for consolidation, leading to uniform and even pressure in all directions. Moreover, this technique has more wastages while removing the laminates of the composites.

Liquid State Composite Manufacturing Processes

The most common methods of Liquid state manufacturing of metal matrix composites are stir casting and squeeze casting. The benefits of the liquid state composite manufacturing techniques are the precise and accurate geometries that can be achieved in low coast and low material wastages. The Liquid state fabrication involves dispersed phase into the liquid molten metal matrix. The fiber is generally used as the dispersed phase, and it increases the bonding strength of the composites. Apart from fibers, ceramic particles and woven fabrics are also used in these composites.

Stir Casting

Stir casting is suitable for manufacturing aluminium matrix composites and hybrid aluminium matrix composite materials [7]. Melting of aluminium is done in a crucible unit and reinforcement material is mixed and stirred using a motorized stirring mechanism to produce even reinforcement composites. The continuously stirred molten metal with reinforcements is poured into the mould cavity to manufacture the composite material with desired geometry. This technique is widely used in the manufacturing of composite materials on large scales.

Squeeze Casting

Squeeze casting is often called a liquid forging technique, in which the composite materials are mixed and poured into the mould cavity [8]. The lower die and upper die clearance lead to squeezing the molten metal, and reinforcement materials tend to fill the space between the lower and upper die [9]. After solidification, the composite material is extracted using this squeeze casting. Stir casting and squeeze casting is exhibited in Fig. (**5**), and Fig. (**6**), respectively.

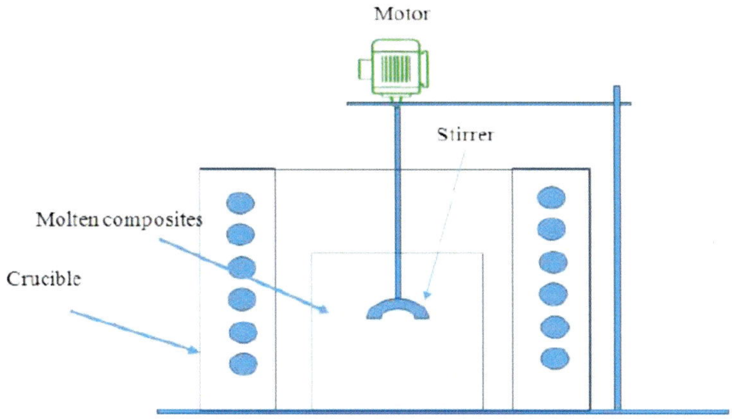

Fig. (5). Schematic diagram of stir casting of composite materials.

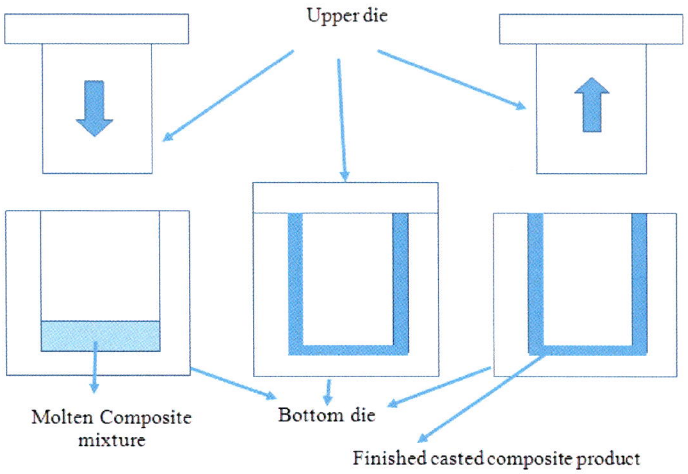

Fig. (6). Schematic diagram of squeeze casting of composite material.

APPLICATIONS OF METAL MATRIX COMPOSITES

In the automotive sector, refractory wool metal matrix components have been explored for engine components. The result of metal matrix composite engine components leads to a reduction of weight and generates less noise and vibrations. Commercially, Toyota used MMC pistons in diesel engines. In aircraft, MMC plays a significant role in manufacturing turbines that can increase thrust efficiency and withstand higher temperatures. In defence, MMC strongly influences the manufacturing of missiles. In inertial guided missiles, MMC replaced beryllium which led to a reduction of weight and manufacturing cost of the missiles. Moreover, MMC is involved in weight reduction in military tanks and guided weapons. MMC strongly influences the manufacturing of electronic goods; due to the lower thermal expansion property of MMC, a higher volume of MMC is used in printed circuit boards and thin-film displays. MMC is widely used in manufacturing sports goods due to its lightweight and good mechanical properties. The components such as bicycle chassis, rim, wheels, golf clubs, horseshoes *etc*., are manufactured using MMC. The manufacturing of Metal matrix composites has a wide scope in the following sectors due to its unique mechanical, thermal, magnetic, and wear resistance properties [10].

1. Aerospace

2. Defence

3. Electronics

4. Sports

5. Biomedical

Due to the more availability of MMC, processing of MMC and manufacturing processes have a wide scope to enhance the optimization of MMC manufacturing processes.

SUMMARY

Composite materials play an essential role in integrating the multifunctional properties of engineering materials. The general manufacturing methods are the solid-state and liquid state manufacturing processes. The metal and ceramic materials are bonded together to form metal matrix composites. The metal matrix composites are used in the manufacturing of satellites, missiles, helicopters, structures, antenna structures, superconductor restraints in fusion power reactors, High-temperature engine parts, Storage battery plates, manufacturing of tool inserts and CNC tool inserts, the base material of silicon carbide is laminated with Titanium nitrides, Zirconium nitrides, *etc*. to improve the tool life efficiency.

CONSENT FOR PUBLICATION

Not applicable.

CONFLICT OF INTEREST

The author declares no conflict of interest, financial or otherwise.

ACKNOWLEDGEMENTS

Declared none.

REFERENCES

[1] K.J. Narayana, and R.G. Burela, "A review of recent research on multifunctional composite materials and structures with their applications", *Mater. Today Proc.,* vol. 5, no. 2, pp. 5580-5590, 2018. [http://dx.doi.org/10.1016/j.matpr.2017.12.149]

[2] Y. Xiao, W. Qiao, H. Fukuda, and H. Hatta, "The effect of embedded devices on structural integrity of composite laminates", *Compos. Struct.,* vol. 153, pp. 21-29, 2016. [http://dx.doi.org/10.1016/j.compstruct.2016.06.007]

[3] C. Calvino, N. Macke, R. Kato, and S.J. Rowan, "Development, processing and applications of bio-sourced cellulose nanocrystal composites", *Prog. Polym. Sci.,* vol. 103, p. 101221, 2020. [http://dx.doi.org/10.1016/j.progpolymsci.2020.101221]

[4] A. Gholampour, and T. Ozbakkaloglu, "A review of natural fiber composites: Properties, modification and processing techniques, characterization, applications", *J. Mater. Sci.,* vol. 55, no. 3, pp. 829-892, 2020. [http://dx.doi.org/10.1007/s10853-019-03990-y]

[5] A. McIlhagger, E. Archer, and R. McIlhagger, "Manufacturing processes for composite materials and components for aerospace applications", In: *Polymer composites in the aerospace industry.* Woodhead Publishing, 2020, pp. 59-81. [http://dx.doi.org/10.1016/B978-0-08-102679-3.00003-4]

[6] K. Palanikumar, and M. Mudhukrishnan, "Technologies in additive manufacturing for fiber reinforced composite materials: a review", *Curr. Opin. Chem. Eng.,* vol. 28, pp. 51-59, 2020. [http://dx.doi.org/10.1016/j.coche.2020.01.001]

[7] S. Ganeshkumar, V. Thirunavukkarasu, R. Sureshkumar, S. Venkatesh, and T. Ramakrishnan, "Investigation of wear behaviour of silicon carbide tool inserts and titanium nitride coated tool inserts in machining of en8 steel", *International Journal of Mechanical Engineering and Technology.,* vol. 10, no. 01, pp. 1862-1873, 2019.

[8] S.G. Kumar, and V. Thirunavukkarasu, "Investigation of Tool Wear and Optimization of Process Parameters in Turning of EN8 and EN 36 Steels", *Asian J. Res. Soc. Sci. Humanit.,* vol. 6, no. 11, pp. 237-243, 2016. [http://dx.doi.org/10.5958/2249-7315.2016.01188.6]

[9] D.D. Soyaslan, "Design and manufacturing of fabric reinforced electromagnetic shielding composite materials", *Tekstil ve Konfeksiyon.,* vol. 30, no. 2, pp. 92-98, 2020. [http://dx.doi.org/10.32710/tekstilvekonfeksiyon.568529]

[10] S. Bahl, "Fiber reinforced metal matrix composites-a review", *Mater. Today Proc.,* vol. 39, pp. 317-323, 2021.
[http://dx.doi.org/10.1016/j.matpr.2020.07.423]

CHAPTER 11

Machining Studies of Basalt Fiber Reinforced Aluminium Metal Matrix Composites Using Abrasive Jet Machining

K. Arunprasath[1,*], P. Amuthakkannan[2], V. Arumugaprabu[3] and **S. Kalirasu[4]**

[1] *Department of Mechanical Engineering, PSN College of Engineering and Technology, Tirunelveli, Tamil Nadu-627152, India*

[2] *Department of Mechanical Engineering, PSR Engineering College, Sivakasi, Tamil Nadu-626140 India*

[3] *Department of Mechanical Engineering, Kalasalingam Academy of Research and Education, Krishnankoil, Tamil Nadu, India – 626126*

[4] *Department of Agriculture Engineering, Sethu Institute of Technology, Pulloor, Kariapatti - 626115, Virudhunagar District, Tamilnadu, India*

Abstract: Aluminium metal matrix composite is widely used in wear resistance applications due to its weight and high strength. It is said to be a perfect composite and alternate material for our day-to-day used supplies. This work aims to study the machining performance of basalt fiber reinforced with aluminium (6082) metal matrix composite. The basalt fiber reinforced aluminium metal matrix was prepared using stir casting techniques with different weights of basalt fiber. Then, the fabricated metal matrix composite was subjected to machining studies using abrasive water jet machining by varying machining parameters such as stand-off distance, travel speed, pressure by measuring the surface roughness, hardness, and density. Furthermore, kerf angle was also measured to find out the optimum weight percentage of the basalt fiber to be reinforced to aluminum metal matrix composite; 2.5 wt % shows the better machining properties while comparing all weight percentages of basalt fiber. Vickers hardness test and density test were also carried out to show the fiber content to be used to reinforce it to the aluminium 6028.

Keywords: Abrasive Jet machining, KERF angle, Metal matrix composites, Stir Casting, Vickers hardness.

* **Corresponding author K. Arunprasath:** Department of Mechanical Engineering, PSN College of Engineering and Technology, Tirunelveli, Tamil Nadu-627152, India; E-mail: aruncmr12@gmail.com

C. Samson Jerold Samuel, M. Suresh & S. Gnanasekaran (Eds.)

INTRODUCTION

Aluminium metal matrix composite is widely used in wear resistance applications due to its weight and high strength. This work aims to study the machining performance of basalt fiber reinforced aluminium (6082) metal matrix composite. This basalt fiber reinforced aluminium metal matrix was fabricated using stir casting techniques with different weights of reinforced basalt fiber, then the fabricated metal matrix composite was subjected to machining studies using abrasive water jet machining by varying machining parameters such as stand-off distance, travel speed, pressure [1]. The measurement of the machining surface of the metal matrix composite (MMC) shows a better material removal rate and surface roughness.This type of intervention is only seen for metal-based composites responsible for providing better strength, wear resistance, low coefficient of thermal expansion, and improved dimensional stability [2]. The metal matrix composite has better strength than fiber contributed polymer matrix composite, but for higher heat resistance, ceramic matrix composite is preferable to other types of composite. Sometimes, due to the dispersion of materials within the composite system, these properties may vary [3]. The microstructure of a particulate that is dispersed in the matrix material is an important factor determining the strength of the composite. The controlled phase structure of the metals will produce optimum results and be suitable to stratify many properties [4]. Now a days, the study on the aluminum metal matrix composite is trending; however, the study can improve with different techniques. This metal has some attractive properties, like good electrical and corrosion resistance, which increases the thirst for research on aluminum matrix composite [5]. A liquid metallurgical technique was followed to prepare the specimen with varied SiC content from 3% of weight to 9% of weight to the aluminum $Al_2 0_3$. During turning, its cutting parameters, namely speed, feed, and depth of cut, are taken as the input parameters, and surface roughness is taken as the response factor.It is inferred that surface roughness will increase if there is an increase in feed rate, and it decreases with an increase in cutting speed [6]. The application of aluminum and its alloy is more widespread, especially in the automobile industry. The engine and its associated parts are made with aluminum. Due to its low density and low weight, it is rapidly used as a core material for many components [7]. Additionally, many of the moving parts of the automobiles and structural surfaces are pre-dominantly replaced with aluminum and its alloys.Aluminum-based composite shows better mechanical and wear properties than other metal composites with less weight as another considerable constraint [8].The rapid development of aluminum 7075 with silicon carbide as composite matrix shows better tribological property than aluminum 6061 alloys. Due to its higher wear rate, hardness, and grip resistance, this alloy composite is superior to other aluminum and some other metal alloys [9]. Numerous tests have been conducted on aluminium metal matrix composite to distinguish the properties of tensile strength and hardness by varying manufacturing techniques. From the performance test conducted on the manufactured gear to the analysis, wear resistance of the aluminium metal matrix was analyzed. The stir casting method is said to be an economical method to produce a very good surface finish on the workpiece material [10]. The improved mechanical properties are seen from fly ash and basalt powder reinforced with Al 6061composites. Aluminum reinforced with fly ash and basalt powder always shows a better agglomeration between the materials, and the stir casting method of manufacturing is one of the notable factors [11]. Basalt powder is a natural material found in volcanic rocks of lava. Three sets of hybrids of aluminum MMC were prepared by varying the weight fraction of the reinforcements as (2.5% of basalt powder + 7.5% of fly ash, 5% of basalt powder + 5% of fly ash, 7.5% of basalt powder + 2.5% of fly ash).The effect of varying reinforcements on the improved mechanical properties such as hardness, tensile strength, and impact strength was noted [12]. The tensile strength of the aluminum (Al) composite and its impact strength are increased with increasing the weight percentage of fly ash, whereas the hardness of the composite is increased with an increasing weight percentage of the basalt powder. From scanning electron microscopic images, the microstructures show very good bonding and dispersion of the reinforcements in the matrix [13]. From the detailed literature review, the combination of aluminum and basalt powder has shown better machining characteristics. The results of some machining studies with abrasive water jet machines are discussed below to fill the research gap on these metal matrix composites.

EXPERIMENTAL WORK

Material Used

The basalt fiber (Fig. **1**) reinforced aluminium metal matrix composite is used in this investigation. The aluminium metal of series 6082 (Fig. **2**) is used in this process. Aluminium material is purchased from the Coimbatore Metal Mart, Coimbatore. The basalt fiber is used as reinforcement material for this metal matrix composite. The basalt fiber is purchased from ASA, TECH Austria.

Fig. (1). Short basalt fiber.

Fig. (2). Aluminium (6082).

Specimen Geometry

The specimen used for this investigation is the basalt fiber reinforced aluminum metal matrix composite.The specimens are categorized into 5 different types based on their fiber weight percentage of 0, 2.5, 5, 7.5, and 10 percent, respectively, for every 500grams of aluminum.

Table 1. Specimen geometry

S.No	Weight of Aluminium(g)	Weight of Basalt fiber (g)	Specimen name
1	500	0	BaAL00
2	500	12.5	BaAL25
3	500	25	BaAL50
4	500	37.5	BaAL75
5	500	50	BaAL100

Preparation of Specimen by Stir Casting Method

Basalt fiber is cut for the required dimension, the pure aluminum rod is cut into 10 pieces and put into a crucible and placed in a casting machine. The casting machine's temperature was set to 750 °C, since it is the melting temperature for aluminum. A cylindrical die is placed over the tray in a horizontal position. The die is made up of cast iron; the die is kept without any disturbance for 2 hours; the crucible is lifted away from the casting machine using the tongs (Fig. **3**). Then, melted aluminum is poured into the cylindrical die, within a few seconds (maximum 12 seconds), the molten aluminum will get solidified (Fig. **4**). After the pouring process, the die is left for 2 hours, and then pure fabricated aluminum is taken out.

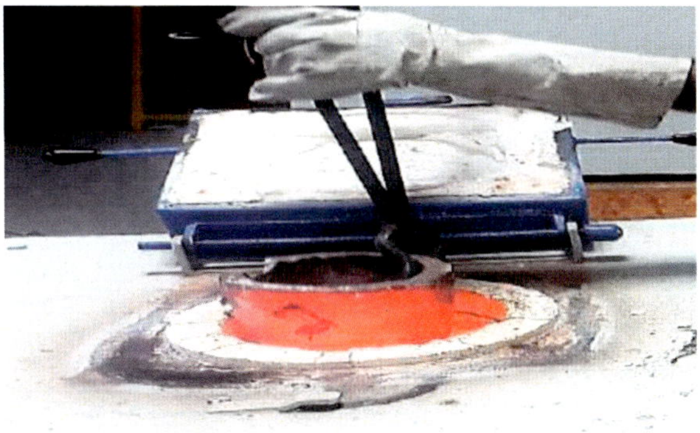

Fig. (3). Fabrication (Casting) process of the aluminum metal matrix composite.

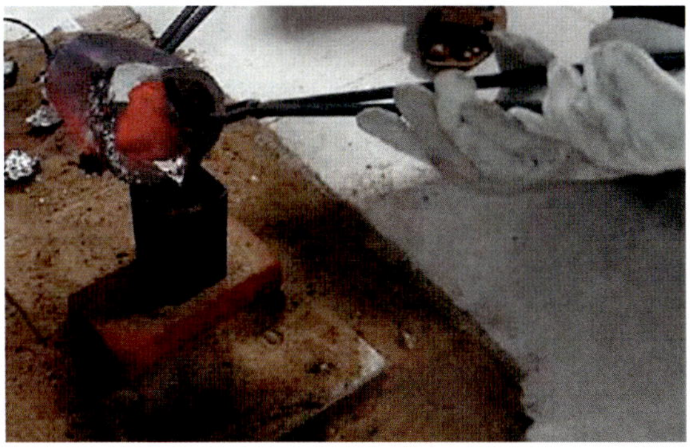

Fig. (4). Fabrication (Casting) process of the aluminum metal matrix composite.

Fig. (5). Cutting of Fabricated Specimen for required dimension.

After the fabrication of pure aluminum, basalt fiber is reinforced with aluminum by varying the weight percentage of basalt fiber to the aluminum. This composition is fabricated, similarly the pure aluminum. Then the stirrer is used to stir the basalt fiber with aluminum, and the fabrication of aluminum matrix composite with 2.5, 5, 7.5, and 10% weight of basalt fiber is processed, and machining is done to cut the fabricated composite for required dimensions (Fig. **5**).

Abrasive Water Jet Machining

To conduct the machining studies on the basalt reinforced aluminium composite, an abrasive water jet machine (Fig. **6**) is taken, and the process parameters of the abrasive water jet machining are optimized for the surface roughness and kerf angle of the machined basalt reinforced aluminium composite.

Fig. (6). Abrasive water jet machine.

The pressure is set in a compressor. The travel speed and the stand-off distance are set in a programmable device. In abrasive water jet machining, the abrasive particle is silicon carbide. Each specimen is cut for a 12mm line to study the various parameters like pressure, travel speed, and stand-off distance. The specimen length is measured using a Microscope. The top and bottom lengths are measured to find out the kerf angle.

RESULT AND DISCUSSION

The machining process is carried out on all the specimens experimentally. From this analysis, we observe that the specimen shows good machining results. The specimens are machined in AWJM with different pressure, stand-off distance, and travel speed to find the kerf angle. The hardness, density, and roughness of the composite are also tested to find the machining properties of the specimen.

Kerf Angle

The comparison of the pressure of water flow, stand-off distance, and travel speed of the nozzle with kerf angle for the specimens of weight percentage 0, 2.5, 5, 7.5, and 10 respectively to find the optimum weight percentage of the basalt reinforced aluminum metal matrix composite which shows the better machining properties with the help of the graph as shown in Figs. (**7**, **8** and **9**)

$$\Theta = \tan\text{-}1((W_t - W_b)/2t) \quad \textbf{(1)}$$

The above formula is used to find out the kerf angle of the aluminum matrix composite, here the W_t is the top kerf width, and W_b bottom kerf width is measured to calculate the kerf angle (Θ) of the specimen, thickness (t) of the specimen is measured to use in the equation to calculate the kerf angle [14]. Kerf angle is calculated and noted for further machining studies.

Fig. (**7**) shows the variation in the kerf angle for each specimen of weight percentage to the pressure of the flow of the water. In this figure, the specimen BaAL75 shows a higher kerf angle in the pressure range of 200 and has a high variation in its kerf angle to pressure. The specimen BaAL100 shows approximately the same range of kerf angle in all ranges of pressures. Fig. (**8**) shows the variation in kerf angle for the specimens of weight percentage for a travel speed of nozzle speed [14]. This specimen BaAL75 shows a higher kerf angle in the Travel speed range of 30. In the range of travel speed of 50, all the specimens show a minimum kerf angle. Fig. (**9**) shows the variation in kerf angle for each specimen of weight percentage for the stand of the distance between nozzle and the specimen. This specimen BaAL 75 shows a higher kerf angle in the range of 1.5 and has higher variation in its kerf angle to stand of distance [15]. The specimen BaAL shows high variation next to the BaAL 75 specimen.

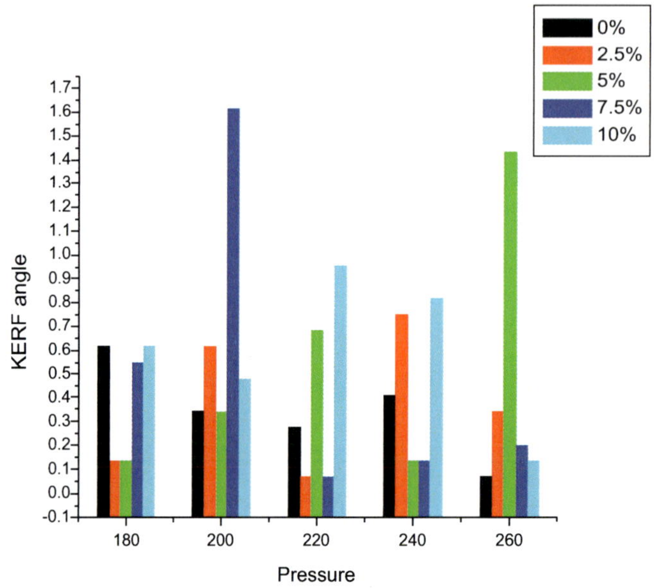

Fig. (7). Pressure *vs*. KERF angle.

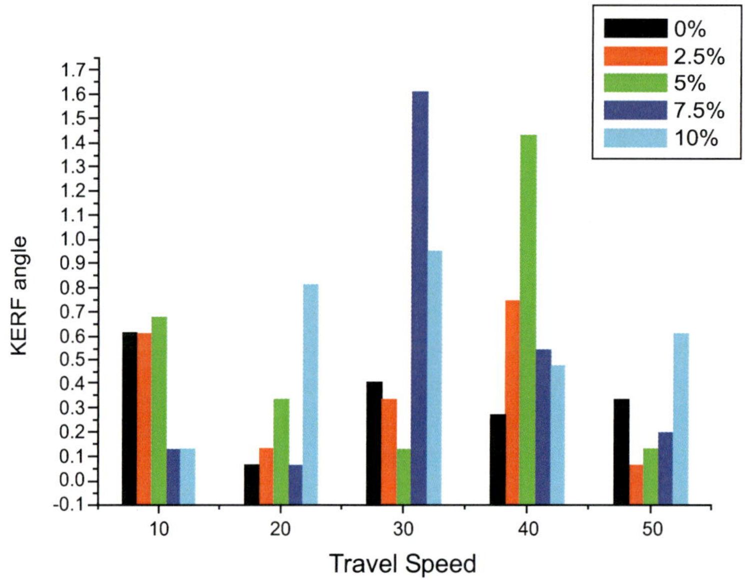

Fig. (8). Travel speed *vs*. KERF angle.

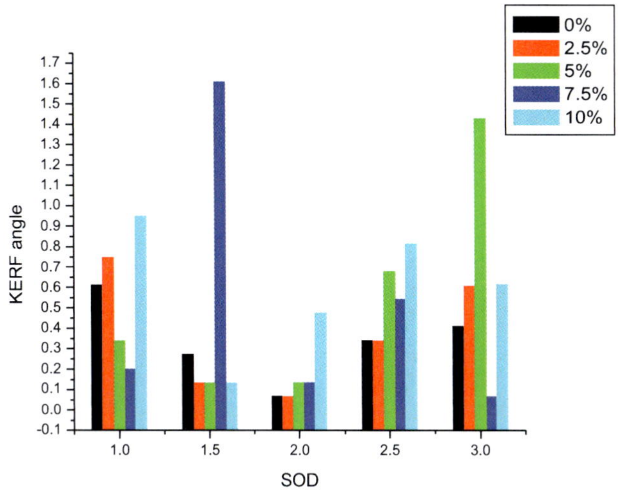

Fig. (9). Stand-off distance *vs*. KERF angle

Surface Roughness

The surface roughness of the specimen is measured using a surface roughness tester. The value is noted for the machining studies Here, surface roughness is the average roughness value for all the specimens. The surface roughness values are plotted as the graph for the percentage variation of basalt in the metal matrix composite specimen. Figs. (**10** - **12**) show the machining performance of the test samples for their average surface roughness values.

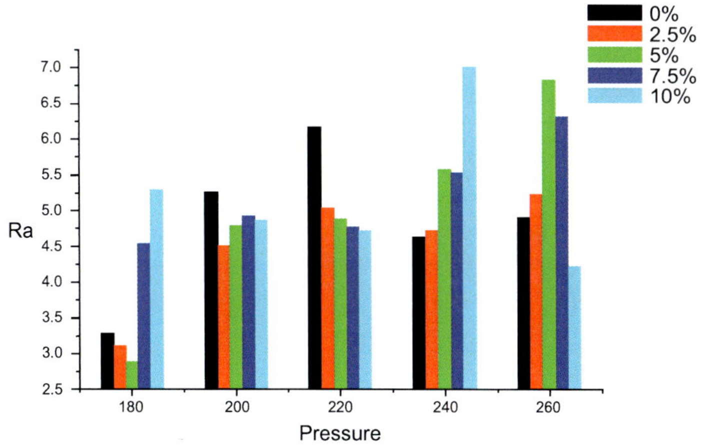

Fig. (10). Ra *vs*. Pressure.

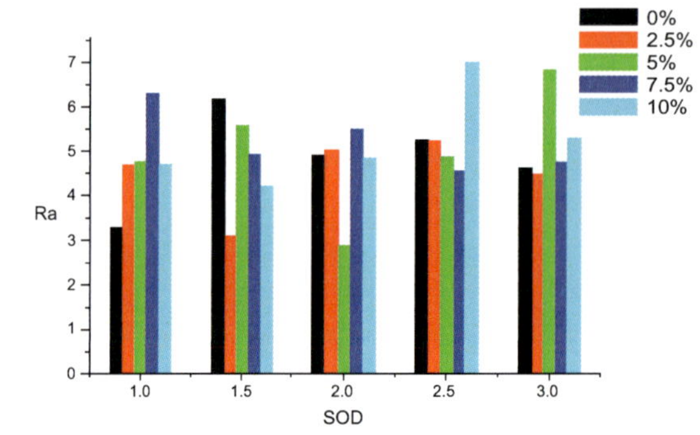

Fig. (11). Ra *vs*. SOD.

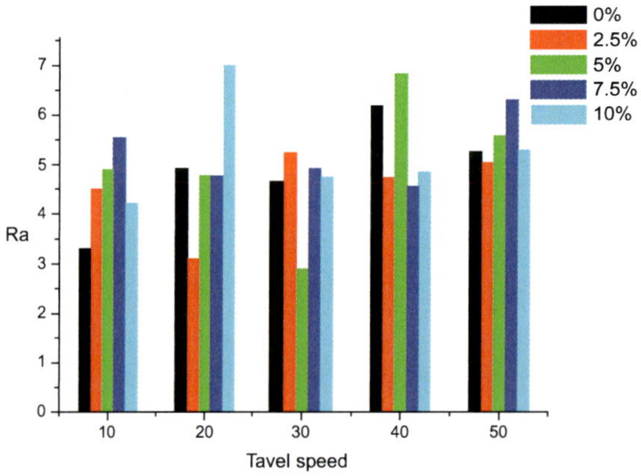

Fig. (12). Ra *vs*. Travel speed.

Taguchi's Main effect plot for Kerf Angle and Surface Roughness

The Taguchi optimization technique is performed using MINITAB software to find the optimum machining parameter and the suitable percentage of basalt.The main effect plot is generated by giving the input and output parameters [16, 17]. Here, L_{25} orthogonal array is selected for the experiment since it has 4 input parameters. Both the output parameters kerf angle and surface roughness are measured for each experiment, and this output is fed into the software for generating the main effect plot, with which the optimum machining parameters and the suitable basalt percentage are inferred [18].

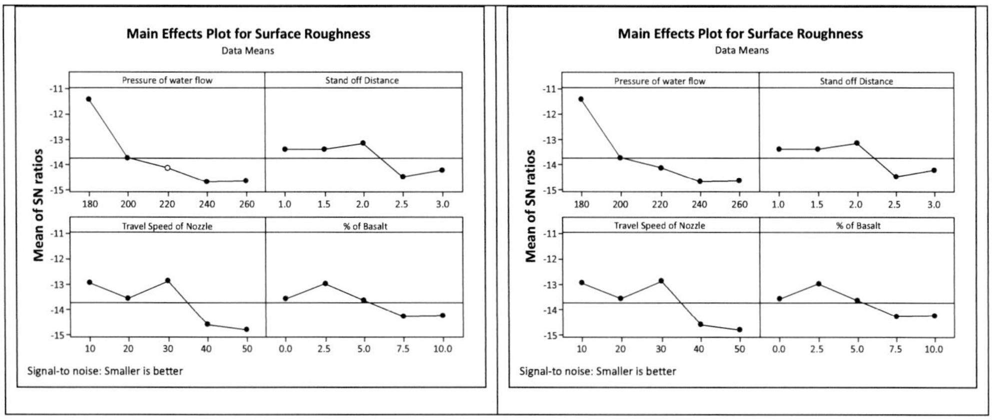

Fig. (13). Main effect plot showing the optimum machining parameters and suitable basalt content.

From Figs. (**8**, the main effect plot generated from MINITAB software, according to the Taguchi optimization technique, the optimum machining parameter for kerf angle is 220kgf/cm^2 Pressure of water flow, 2mm Stand-off distance, 20mm/s travel speed. Moreover, the suitable basalt percentage is 2.5 wt. %. The optimum machining parameter for surface roughness is 180kgf/cm^2 pressure of water flow, 2mm Stand-off distance, 30mm/sec travel speed, and the suitable basalt percentage is 2.5 wt. %.

Hardness Test

Table 2. Tabulation Vickers hardness test

S. No	% of Basalt	Load (N)	Least Count	Impression	V.H.N
1	0	50	2	0.24	1557
2	2.5	50	2	0.37	674
3	5	50	2	0.44	475
4	7.5	50	2	0.49	384
5	10	50	2	0.55	304

The hardness of the prepared aluminum matrix composite specimens is measured using Vicker's hardness tester. The hardness values are noted for the performance check of the prepared metal matrix composite [19, 20].

Density Test

The test shows that the density is decreased when the fiber content goes beyond the 7.5 weight percentage. It is noted that for 10 weight percentage of fiber, the density of 1.159 g/cm^3

Table 3. Tabulation of density for different fiber content.

S. No	Fiber Percentage	Density(g/cm^3)
1	0	2.644
2	2.5	1.171
3	5	1.165
4	7.5	1.168
5	10	1.159

CONCLUSION

Thus, the basalt fiber reinforced aluminium metal matrix composite specimen was prepared by the stir casting method. The following conclusions were drawn for the specimen with a different weight percentage of basalt fiber 0 wt. %, 2.5 wt. %, 5 wt. %, 7.5 wt. %, 10 wt. %, mixed with aluminium (6082) and prepared by a stir casting machine. The machining process was conducted using Abrasive Water Jet Machining (AWJM) for the specimen.

- Kerf angle and average surface roughness are measured for all the experiments with different combinations of pressure, stand-off distance, travel speed, and the weight percentage of fiber, and the experiment table is designed according to the orthogonal L_{25} array.
- From the main effect plot generated using Taguchi analysis with the statistical tool MINITAB, it is found that 2.5 wt. % of basalt fiber gives better surface roughness and kerf angle.
- The optimum machining parameter from Taguchi analysis is 220kgf/cm^2 pressure of water flow, 2mm Stand-off distance, 20mm/s travel speed for the kerf angle, and 180kgf/cm^2 pressure of water flow, 2mm Stand-off distance, 30mm/sec travel speed for average surface roughness.
- Vickers hardness test and density test also show that the 2.5 wt. % the weight percentage of basalt fiber gives good machining performance compared to the other fiber content.

CONSENT FOR PUBLICATION

Not applicable.

CONFLICT OF INTEREST

The author declares no conflict of interest, financial or otherwise.

ACKNOWLEDGEMENTS

Declared none.

REFERENCES

[1] T. Sasimurugan, and K. Palanikumar, "Analysis of the machining characteristics on surface roughness of a hybrid aluminum metal matrix composite (Al6061-SiC-Al_2O_3)"., *J. Miner. Mater. Charact. Eng.*, vol. 10, no. 13, pp. 1213-1224, 2011. [http://dx.doi.org/10.4236/jmmce.2011.1013094]

[2] D. Vengatesh, and V. Chandramohan, "Aluminium alloy metal matrix composite: survey paper", *International Journal of Engineering Research and General Science.*, vol. 2, no. 6, pp. 792-796, 2014.

[3] V. Balaji, N. Sateesh, and M.M. Hussain, "Manufacture of aluminum metal matrix composite (Al7075-SiC) by stir casting technique", *Mater. Today Proc.*, vol. 2, no. 4-5, pp. 3403-3408, 2015. [http://dx.doi.org/10.1016/j.matpr.2015.07.315]

[4] K.A. Prasath, P. Amuthakkannan, V. Manikandan, R. Jegadeesan, and M. Selwin, "Novel topological approach in mechanical properties of basalt/flax hybrid composites", In: *In AIP Conference Proceedings* vol. 2128. , 2019, no. 1, p. 020001. AIP Publishing LLC.

[5] S.M. Khalili, V. Daghigh, and R. Eslami Farsani, "Mechanical behavior of basalt fiber-reinforced and basalt fiber metal laminate composites under tensile and bending loads", *J. Reinf. Plast. Compos.*, vol. 30, no. 8, pp. 647-659, 2011. [http://dx.doi.org/10.1177/0731684411398535]

[6] M.H. Rahman, and H.M. Al Rashed, "Characterization of silicon carbide reinforced aluminum matrix composites", *Procedia Eng.*, vol. 90, pp. 103-109, 2014. [http://dx.doi.org/10.1016/j.proeng.2014.11.821]

[7] R.D. Johnson, K.L. Wins, A. Raj, and B.A. Beatrice, "Optimization of cutting parameters and fluid application parameters during turning of OHNS steel", *Procedia Eng.*, vol. 97, pp. 172-177, 2014. [http://dx.doi.org/10.1016/j.proeng.2014.12.239]

[8] R. Johnson, and Joel Deepak , "Influence of drilling parameters in drilling of Hamc with vegetable based cutting fluid", *Adv. Nat. Appl. Sci*, vol. 10, pp. 300-306, 2016.

[9] V. Daghigh, T.E. Lacy Jr, H. Daghigh, G. Gu, K.T. Baghaei, M.F. Horstemeyer, and C.U. Pittman Jr, "Heat deflection temperatures of bio-nano-composites using experiments and machine learning predictions", *Mater. Today Commun.*, vol. 22, 2020.100789 [http://dx.doi.org/10.1016/j.mtcomm.2019.100789]

[10] R. Butola, S. Kanwar, L. Tyagi, R.M. Singari, and M. Tyagi, "Optimizing the machining variables in CNC turning of aluminum based hybrid metal matrix composites", *SN Applied Sciences.*, vol. 2, no. 8, pp. 1-9, 2020. [http://dx.doi.org/10.1007/s42452-020-3155-8]

[11] K.R. Sumesh, and K. Kanthavel, "Abrasive water jet machining of Sisal/Pineapple epoxy hybrid composites with the addition of various fly ash filler", *Mater. Res. Express,* vol. 7, no. 3, 2020.035303 [http://dx.doi.org/10.1088/2053-1591/ab7865]

[12] M. Lamea, V. Daghigh, M. Soroush, and K. Nikbin, "The Buckling Behavior of Vacuum-Infused Open-Hole Unidirectional Basalt-Fiber Composites Experimental and Numerical Investigations", *Mech. Compos. Mater.,* vol. 55, no. 6, pp. 761-774, 2020.
[http://dx.doi.org/10.1007/s11029-020-09848-8]

[13] P.R. Doddi, R. Chanamala, and S.P. Dora, "Effect of fiber orientation on dynamic mechanical properties of PALF hybridized with basalt reinforced epoxy composites", *Mater. Res. Express,* vol. 7, no. 1, 2020.015329
[http://dx.doi.org/10.1088/2053-1591/ab6771]

[14] A. Jain, and B. Singh, "Parametric Analysis during Laser Cutting of Basalt–Glass Hybrid Composite", *Lasers in Manufacturing and Materials Processing.,* vol. 7, no. 1, pp. 111-139, 2020.
[http://dx.doi.org/10.1007/s40516-019-00110-1]

[15] M. Bulut, O.Y. Bozkurt, A. Erkliğ, H. Yaykaşlı, and O. Ozbek, "Mechanical and Dynamic Properties of Basalt Fiber-Reinforced Composites with Nanoclay Particles", *Arab. J. Sci. Eng.,* vol. 45, no. 2, pp. 1017-1033, 2020.
[http://dx.doi.org/10.1007/s13369-019-04226-6]

[16] V. Sharma, M.L. Meena, M. Kumar, and A. Patnaik, "Mechanical and three-body abrasive wear behavior analysis of glass and basalt fiber-reinforced epoxy composites", *Polym. Compos.,* vol. 6, no. 11, pp. 121-130, 2020.
[http://dx.doi.org/10.1002/pc.25670]

[17] M.G. Sohail, W. Alnahhal, A. Taha, and K. Abdelaal, "Sustainable alternative aggregates: Characterization and influence on the mechanical behavior of basalt fiber reinforced concrete", *Constr. Build. Mater.,* vol. 255, 2020.119365
[http://dx.doi.org/10.1016/j.conbuildmat.2020.119365]

[18] S. Ilangovan, S.S. Kumaran, and K. Naresh, "Effect of nanoparticles loading on free vibration response of epoxy and filament winding basalt/epoxy and E-glass/epoxy composite tubes: experimental, analytical and numerical investigations", *Mater. Res. Express,* vol. 7, no. 2, 2020.025007
[http://dx.doi.org/10.1088/2053-1591/ab6e36]

[19] C. Sivakandhan, G.B. Loganathan, G. Murali, P.S. Prabhu, S. Marichamy, G.S. Krishnan, and R. Pradhan, "Material characterization and unconventional machining on synthesized Niobium metal matrix", *Mater. Res. Express,* vol. 7, no. 1, 2020.015018
[http://dx.doi.org/10.1088/2053-1591/ab624d]

[20] F.A. Arpatappeh, M.A. Azghan, and R. Eslami-Farsani, "The effect of stacking sequence of basalt and Kevlar fibers on the Charpy impact behavior of hybrid composites and fiber metal laminates", *Proc. Inst. Mech. Eng., C J. Mech. Eng. Sci.,* vol. 25, no. 16, 2020.0954406220914325
[http://dx.doi.org/10.1177/0954406220914325]

CHAPTER 12

Multi-objective Optimization on Processing Parameters for Micro ECM of Inconel Alloy Using Taguchi Based Grey Relation Analysis

Selvan T. A.[1,*], **A. Viswanathan**[2], **S. Madhankumar**[3] and **K. M. Senthilkumar**[4]

[1] *Department of Mechatronics Engineering, Sri Krishna College of Engineering and Technology, Coimbatore, India-641008*

[2] *Department of Mathematics, SNS College of Technology, Coimbatore, India – 641035*

[3] *Department of Mechatronics Engineering, Sri Krishna College of Engineering and Technology, Coimbatore, India-641008*

[4] *Department of Mechanical Engineering, Kumaraguru College of Technology, Coimbatore, India-641049*

Abstract: This study analyzed the results of Inconel 718 alloy on substance extraction capacity, including surface roughness. The analysis was carried out in micro electro-chemical machining by applying various variables and employing stainless rod material as a substrate. The optimizing method depended on Taguchi's methodology-based grey relation technique using Minitab 19 tools. Predictor variables were designed for the L9 dimensional set using Taguchi's model with 3 input variables: voltage, electrolyte concentration, and current. Taguchi's model was employed to estimate the effect of input variables upon the rate of dissolution and surface roughness of the substance. For predictions, regression methodologies have been introduced. In the end, the optimal run for the substance removal capacity was 16 volts, 0.40 mole/lit, and 0.8 amp. The optimal run for surface roughness was 8 volts, 0.40 mole/lit, and 1 amp. The combined optimal run was achieved in run number 3. Subsequently, a confirmatory test was carried out to verify the experimental results.

Keywords: Inconel 718 alloy, Grey relation optimization, Regression, Taguchi's model.

INTRODUCTION

It has been widely recognized that the results of the processing variables for a specified machining process cannot be decided at choice; optimization strategies

* **Corresponding author Selvan T.A.:** Professor and Head, Department of Mechatronics Engineering, Sri Krishna College of Engineering and Technology, Coimbatore, India-641008; E-mail: selvan@skcet.ac.in

C. Samson Jerold Samuel, M. Suresh & S. Gnanasekaran (Eds.)

must be used to determine the optimal quality of the control variables in order to make the best usage of resources such as energy, duration, and substances [1]. Micro electro-chemical machining (μ-ECM) is indeed an unconventional manufacturing technology, which been recognized for its various machine components [2,3]. Previously the drilling of complicated prototypes was challenging. Consequently, with the introduction of revolutionary metalworking technologies integrating chemicals and mechanical/electrical methods, every production line has been revolutionized. New materials were found with a strength-to-weight thermal expansion, toughness, and even complicated structures that might require higher consistency [4,5]. Advanced and innovative drilling processes sometimes allude to untraditional machining operations that strip substances without thermal energy. However, all forms of materials could be manufactured through this method. Through extremely dynamic-precision and time-sensitive situations, μ-ECM have a variety of applications [6]. Quite explicitly, μ-ECM is a procedure based on anodized regulated dissolve of a work-piece anode to use the apparatus as an electrode inside an electrode surface [7], [8]. The electrolyte streams in between electrodes take the soluble material down. When Gusseff developed the first ECM in 1929, their industrial uses were already applied to electro-chemical machining, brazing, brushing as well as scrubbing. Importantly, ECM reports were noticed to be much more favorable for maximum strength alloys. Currently, ECM has been widely recognized for some of its metalworking ability, whereas the quality of its machining model profile is in consideration [9]. The electric arc produced by electrolyte removed the material from the workpiece [10]. The substrate is being deteriorated in order to comply with Faraday's electrodeposition law. μ-ECM techniques become commonly applicable in fields including the aircraft and manufacturing sectors for producing and end products of a range of components with a radius of a few micrometers. Electrolyte concentration, current, and voltage are among the variables mostly deemed. While there is a likelihood of enhancing the validity of the research, the reliance of consistency on varying factors requires a detailed study to find the cause and effect of the various variables [11].

Throughout the context from the above details, this work was conducted to determine the optimal opportunities of various processing variables in order to improve the precision of the μ-ECM operation. The mathematical equations have been developed to predict the material removal rate and surface roughness. The analysis foresaw scientific evidence from the experimentation conducted to evaluate the impact of design to operate variable uncertainties mostly on MRR and SR of the Inconel 718 alloy.

MATERIAL AND METHODS

Working Principle of Micro ECM

The μ-ECM is determined by the method of electrolysis. This method works in the existence of an electrolyte stream that is pumped, employing an appropriate pumping filled with the interlayer distance in between anode (work) and the negative electrode (tool) [12]. The electrolyte is used as a conductive medium between workpiece and tool. However, as micromachining aims to extract the material, the substrate is attached to the anode and the electrode to the cathode [13]. The material is extracted by the regulated dissolution of a positive terminal as per the best-known law of electrolysis of Faradays [14]. Usually, a favorable sodium chloride salt (NaCl) is used as an electrolytic solution. The μ-ECM device concept and operation particulars are seen in Fig. (**1**).

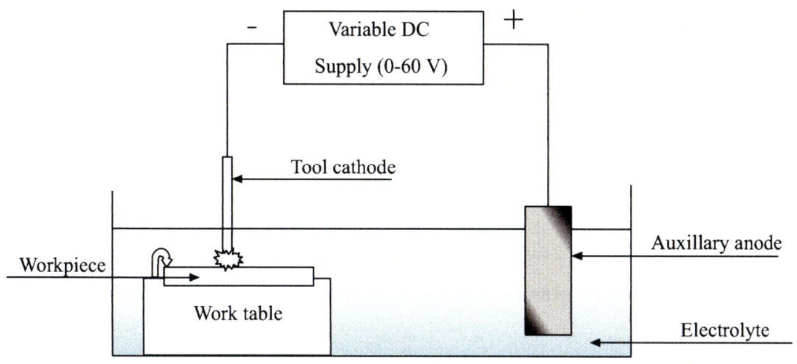

Fig. (1). The μ-ECM device concept and operation particulars.

The servo controller regulates the movement of an electrode according to the work material and can pursue in the chosen direction. Thus, it regulates the breadth of the distance within a limit so that th release phase can begin. Unless the cathode acceleration of the device is too rapid and hits the work-piece, a high voltage occurs [15]. Small explosion contributes in less metal removal since the voltage drop seen between the cathodes is smaller and the current is constrained by the motor. If the drill interface advances gradually, the distance becomes too large, and the electric current rarely occurs. A further feature, its servo device, would be to remove even the instrument electrodes whenever the hole position is damaged. The breadth cannot be determined during drilling other probability distributions that are important for servo regulation. Fig. (**2**) displays the line representation of the configuration of the μ-ECM.

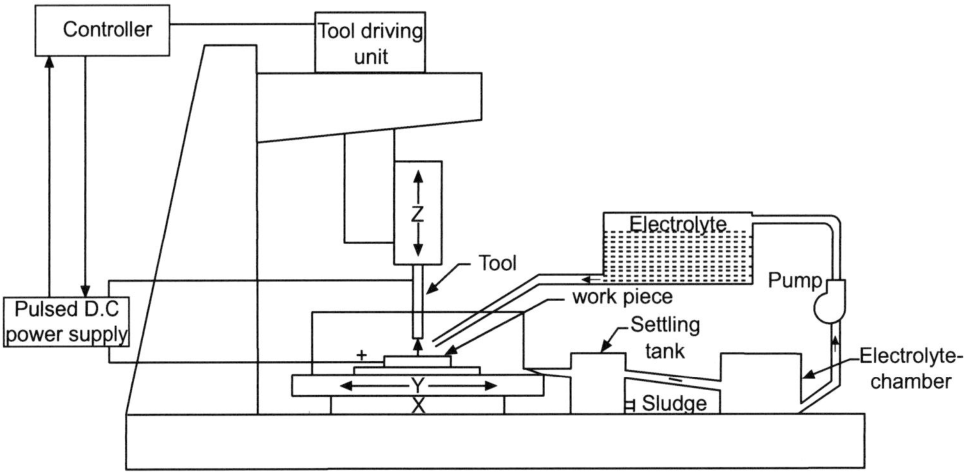

Fig. (2). The line representation of the configuration of the μ-ECM.

In the circumstance of the μ-ECM operation, the distance of around 0.01 to 0.075 mm between the instrument and the work material shall be maintained. In order to minimize the distance, the electromotive force between some of the machines and the job is minimum, and the pressure is optimum, thus, the entire material is eliminated. The material is plunged into the substance based on how quickly the material ought to be extracted [16,17]. The flow of the drill tumble is regulated by a hydraulic system providing the same amount of fed rate. For this experimentation, the selected feed rate was 0.04 m/min. The electrolyte is important for the activity of the electrolyte phase. The electrolyte in the micro ECM conducts the following three fundamental functions:

- Fulfilling the electrical circuit and enabling the broad current to flow through.
- Maintaining the chemical processes needed;
- Eliminating the energy created as well as the waste materials.

The chloride (electrolyte solution) heat should be stable such that there is no variance of dielectric constant. Unless the heat of the electrolyte becomes small, this implies a lower rate of removing contaminants because if the temperature goes up, the chlorine ions may be vaporized. The chloride temperatures should therefore be preserved around 25°C and 60°C. The electron mobility of its chloride solution should be strong as well as the volatility and degradation of the fluids be minimal. The chlorine ions are injected at roughly 1.4 N/m^2 and a velocity below 25 m/s. For nickel-based alloys, the electrolyte used is NaCl and a blend of brine and H_2SO_4.

Selection of Material

Inconel 718 alloy was chosen as an investigational work material. Taguchi's pattern (L9) is designed for the Inconel 718 specimen, and the substrate has a span of 4 cm, breath of 4 cm, and thick of 0.04 cm. The chemical components of the Inconel718 alloy substance are presented in Table **1** as follows [18, 19]. The mechanical properties of Inconel 718 alloy are mentioned in Table **2** [20].

Table 1. Chemical components of Inconel718 alloy.

Element	Al	Ni	Si	Fe	Mn	Ti	C	Co	Mb	Cb	Cr
% Composition	0.8	52.82	0.35	17	0.35	0.6	0.08	1	3	5	19

Table 2. Mechanical properties of Inconel718 alloy.

Property	Ultimate tensile strength (MPa)	Tensile yield strength (MPa)	Hardness number (HRC)
Value	1325	1100.5	42.5

Response Variables

For this investigation, the variables current, voltage and electrolyte concentrations are selected as the input variables. The output responses variables are material removal rate (MRR) and surface roughness (SR). Electrolysis is the foundation for the extraction of substances. Faraday presented two rules for this; the first was that the quantity of chemical change caused by electrical current, which would be the quantity of certain material subsided or stored, is equivalent to the quantity of electrical energy transmitted. Moreover, "The multitude of distinct components disbanded *via* an equitable cost of heat is approximately equal to one's chlorine corresponding lifts". The MRR factor is a measure including its rotational speed that governs the flow of electrons between the job as well as the technique [21]. The load resistance decreased when the air flows towards its machined surface. As a result, a stable separation among tool - work has been identified. It should be observed that the highest rotational speed is indeed efficient and provides the highest substrate finishing efficiency. That being said, the feeding rate has been reduced by eliminating heat energy and work-piece materials. The MRR is slowed by lower voltage, low chloride content, and low thermal conductivity. The greatest benefit of its material extraction process is that it doesn't induce any unintended crystalline phase in traditional techniques. The key advantages are stress-free metal casting, stamper-free structures, decreased tool life, and elimination from toxic effects to a machined surface. Such methods have little effect on the properties like yield stress, ultimate tensile stress, toughness, and so on. μ-ECM might generate an SR around 0.50 μm through turning the tool or through functioning. Every distortion on the cutting tool creates a copy of the work object. Consequently, the tool exterior should be refined. The performance is excellent in a rugged substance. For optimal tool life, conscientious substrate layout, feed flow rate, voltage, current, including surface-enhancing chloride compounds are chosen.

Due to the convenience described herein, μ-ECM is used in different industries. Drilling variables and perhaps other considerations specifically verify the result and quality of the machine tool. It then requires one to find the optimum conditions for achieving improved output values together with the economy of the system [22]. Numerous studies were conducted using μ-ECMs of various metals, including implementing multiple methods to seek an optimum value. The electro-chemical machining of predominant Ti, Ni-based alloys and its use in machining aero motor parts have been explored [23]. ECM provides an important alternate solution manufacturing different complicated molded parts such as molds, turbine parts, and micro-dimensional parts. The μ-ECM procedure for varying electrolyte blends has been examined and it has been found that low levels would result in greater precision [24]. The μ-ECM method for Ni-based materials has been investigated. It has been demonstrated that the production of a performance commodity can be accomplished by following the same degradation of molecular operation [25]. The μ-ECM of cylinder-Steels using different electrolyte solutions have been explained [26]. It is suggested that its electrodes' rotational speed influences the MRR. The particular features of its μ-ECM machining were also investigated, such as the electrode's nose radius, the volts, the chloride pressure drop in order to evaluate the absolute best output measurement processing parameters, including MRR, tool wear, spatial precision, and system expense [27]. The option of process variables to decide the absolute best metalworking efficiency is still difficult. It also formulated the optimization technique of Ti alloy chemical reactions drills using the grey-based Taguchi process [28]. The EDM of different Ni-based metals, Inconel 625 and 718, was investigated, respectively, using Taguchi layout strategy to evaluate optimum performance parameters [29]. Multi - objective-making method grey relation technique was used for end milling of metals [30]. The Taguchi-based grey methodology was used to solve the optimization problems of this method. The output responses considered in this analysis are MRR and SR.

Design of Experiment and Grey Relation Technique

Taguchi's theory was formulated by Dr. Genichi Taguchi. It is an essential guide for implementing a high-quality production method [31]. Taguchi's Orthogonal Array (OA) includes a suite of tests with such a reduced set of possibilities, and

Taguchi's signal-to-noise ratios (S/N), which are exponential features of the target outputs and assist as decision variables throughout the optimization problem [32]. Taguchi methodology uses a mathematical efficiency indicator named a signal-t--noise ratio. The S/N ratio reflects both the median and the uncertainty. The amount S/N is the proportion of the mean (Signal) to the confidence interval (Noise). The risk factors on the consistency of the product/process are also to be optimized. The normal S/N ratios commonly shown are Nominal is Best (NB), lower is Better (LB), and higher is better (HB). The normalized value is between 0 and 1. Zero is perhaps the worst standard to be refused, and one is the most acceptable. Since the S/N ratio is represented as a mean, the signal to the noise reflects a divergence from the goal. Trying to maximize the ratio S/N specifies the overall variance and that the ratio S/N is to be maximized. The Taguchi approach aims to minimize noise since the reduction of the noise element is not feasible. The Taguchi approach gives many lightweight variations for studies with the optimal configuration of control systems variables. This is why Taguchi methods are used to design experiments with stochastic optimization techniques to achieve the goal. The selected L9 orthogonal array (3 factors and 3 variables) is presented in Table **3**.

Table 3. Design of experiment using Taguchi's design.

Run	Voltage	Electrolyte Concentration	Current
1	1	1	1
2	1	2	2
3	1	3	3
4	2	1	2
5	2	2	3
6	2	3	1
7	3	1	3
8	3	2	1
9	3	3	2

MINITAB'19 provides both simple and complex approaches for testing static response measures. The normal distribution of importance has a defined standard. In transient behavior tests, the skill acquisition works across a spectrum of variables, and the purpose is to enhance the interaction between the feedback signal and the output response. MINITAB measures response scales and produces significant barrier and relationship plots for - S/N ratios to indicate reliability vs. process parameters. It also implies (static design) or slope (dynamic design) *vs.* dependent variables [21]. The Taguchi model and OA are the process whereby various styles are used *i.e.*, two, three, four, five, and a mixed form stage. First of all, the decision was taken about how many variables were appropriate for the study. Afterward, three voltage, electrolyte concentration, and current parameters were assessed throughout this test. Table **4** shows machining parameters and their levels.

Table 4. Input process variables and their levels.

Process Variables (unit)	Level 1	Level 2	Level 3
Voltage (volts)	8	12	16
Electrolyte Concentration (mole/lit)	0.22	0.31	0.40
Current (amps)	0.6	0.8	1

The objective function is used to represent the relative reference of each model [33]. In other terms, the key factor approach is used to evaluate the appropriate weighting values of each performance characteristic, where this process is paired with a grey hierarchical approach and used in problems with multiple performance characteristics, representing the possible influence of each objective function [34,35]. The measures taken in the Grey relation technique combined with the key measurement model are all as below: -

Step 1: - Modify experimental results to S/N quantities.

Step 2: - The modified value of S/N is normalized.

$$Higher\ the\ better, A_{pq} = \frac{Cpq - Min(Cpq)}{Max(Cpq) - Min(Cpq)} \quad (1)$$

$$Lower\ the\ better, A_{il} = \frac{Max(Cpq) - Cpq}{Max(Cpq) - Min(Cpq)} \quad (2)$$

Step 3: - The related grey relation coefficients are determined after S/N quantity normalization.

$$Grey\ relation\ coefficient, \theta pq = \frac{Min(\emptyset pq) + (0.5 \times Max(\emptyset pq))}{\emptyset pq + (0.5 \times \max(\emptyset pq))} \quad (3)$$

Step 4: - Measure the grey relation grade through using principal component model.

$$Overall\ grey\ relation\ grade, \gamma j = \frac{\sum_{i=1}^{n} \theta pq}{n} \tag{4}$$

Step 5: - Now conduct a mathematical model of the variability to achieve an optimum value of the operating conditions.

RESULTS AND DISCUSSION

Experimental Results

Before conducting the study, the initial weight of a work material must be calculated to determine the MRR. After establishing all criteria in the command prompt, such as the feed rate, the voltage, and the work-piece in the tank, machining starts with the use of stainless-steel electrodes. Throughout this phase, the work-piece is held on a flat surface and, using the electrodes; it is held in an upright direction, starting from the center position. Precaution must be taken to ensure that the edge of the electrodes does not contact the surface of the material. The duration of milling of the material surface at some pressure, chloride concentration and voltage are observed throughout the process. Nine tests were performed using an OA of L9 by adapting the three levels in each of the three factors, like the voltage, concentration, and current, as seen in Table **5**, for which the values MRR and SR were assessed. The MRR is determined by the equation (5). The original and final weighs of the work material are determined by an automated measuring system with a precision of 0.001 g. The roughness of machines' work-pieces is reported by utilizing the SRG4500 hardness test machine. The measured results of MRR and SR are included in Table **5**. Fig. (**3**) shows the work-pieces after machining in micro electrochemical machining.

Table 5. Experimental results of MRR and SR.

Run	Voltage (A) (volt)	Electrolyte concentration (B) (mole/lit)	Current (C) (amps)	Material Removal Rate (mm^3/min)	Surface Roughness (µm)
1	8	0.22	0.6	0.0365	0.5582
2	8	0.31	0.8	0.0389	0.4932
3	8	0.40	1	0.0891	0.3854
4	12	0.22	0.8	0.0404	0.6151
5	12	0.31	1	0.0785	0.5299
6	12	0.40	0.6	0.0796	0.4963
7	16	0.22	1	0.0733	0.6452
8	16	0.31	0.6	0.0562	0.6435

(Table 5) cont.....

Run	Voltage (A) (volt)	Electrolyte concentration (B) (mole/lit)	Current (C) (amps)	Material Removal Rate (mm^3/min)	Surface Roughness (μm)
9	16	0.40	0.8	0.0911	0.5448

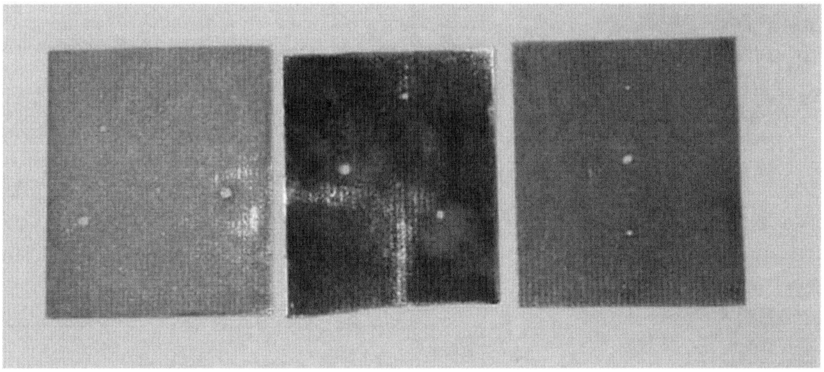

Fig. (3). Inconel 718 alloy after micro electro-chemical machining.

$$Material\ removal\ rate = \frac{Original\ weight - Final\ weight}{Density \times Machining\ time} \quad (5)$$

From these measurements, the output responses of Inconel 718 alloys show that they agree with input processing parameters. The larger MRR is achieved at run number 9 (16 volts, 0.40 mole/lit, and 0.8 amp) of 0.0911 mm^3/min, and the better (lower) SR is got at run number 3 (8 volts, 0.40 mole/lit, and 1 amp) of 0.3854 μm. But we cannot predict the optimum processing variables in combination of MRR and SR. This was found by using the grey relation technique and analysis of Taguchi design.

Influencing of μ-ECM Input Process Variables on MRR and SR

Taguchi's experiment model uses fewer runs to test the influence of process variables. Taguchi is a simple, time-saving approach and could usually be extended to any technical challenges. Tables **(6a and b)** present the response table for MRR and SR, with respect to input processing variables. Those response tables will also show which operating variables have a stronger impact, mostly on reactions calculated by assigning the model variable a level. Often, the optimum value is being derived from the response set. The maximum priority related to the individual standing in the response tables is the estimated concentration for MRR and the lowest mean for the SR. The responsible table for means explains that the effect of influences of processing variables varies for MRR and SR. For MRR the electrolyte concentration was the most influencing variable, followed by current and then voltage. But for SR, the voltage is the most influencing variable, followed by voltage and then-current.

Table 6 (a). Response table for means of MRR.

Level	Voltage (volt)	Electrolyte concentration (mole/lit)	Current (amps)
1	0.05483	0.05007	0.05743
2	0.06617	0.05787	0.05680
3	0.07353	0.08660	0.08030
Delta	0.01870	0.03653	0.02350
Rank	3	1	2

Table 6 (b). Response table for means of SR.

Run	Voltage (volt)	Electrolyte concentration (mole/lit)	Current (amps)
1	0.4789	0.6062	0.5660
2	0.5471	0.5555	0.5510
3	0.6112	0.4755	0.5202
Delta	0.1322	0.1307	0.0458
Rank	1	2	3

The key impact chart seems to represent its median or ways of reaction at each step of the component or framework provided. The effect size plot allows to evaluate the effects of the specific input variables on the calculated results by ignoring the impact of all other existing input variables. Its effect size plotlines at each reply have been stated below. Even though MRR in micro μ-ECM is quite poor compared to the traditional manufacturing process, yet it seems to be a favored choice for heavy metal contamination-to-cut substances like nickel-based superalloy, Inconel 718 alloy is part of the same group of challenging and complex feature alloys. The efficiency of the μ-ECM can indeed be calculated using the MRR, so it would be vital to recognize the effect of the mechanical properties mostly on MRR during the μ-ECM of Inconel 718 alloy. The SR of the machining process could be calculated from its rough surface. During the μ-ECM technique, the SR obtained is of small magnitude because no more finishing procedure is needed again for the machined surface.

Fig. (**4a** and **4b**) show the main effect plot for signal-to-noise ratios of MRR and SR, respectively. The graph identifies that the MRR increases when voltage and electrolyte concentration increases. Moreover, there is a fluctuation in the response with respect to input process parameter current. The SR value increases when the input variables, electrolyte concentration and current increases, yet the SR decreases when voltage is going down from 16 to 8.

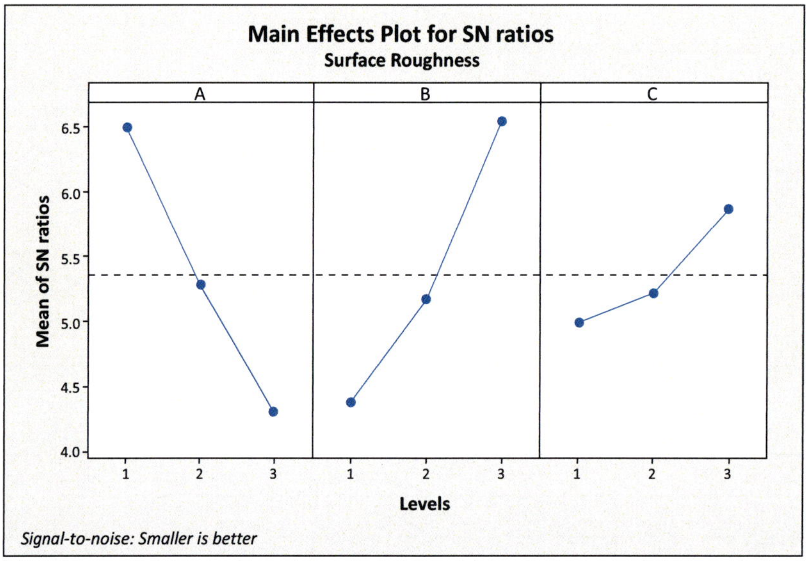

Fig. 4 (a). Main effect plot for SN ratios of MRR.

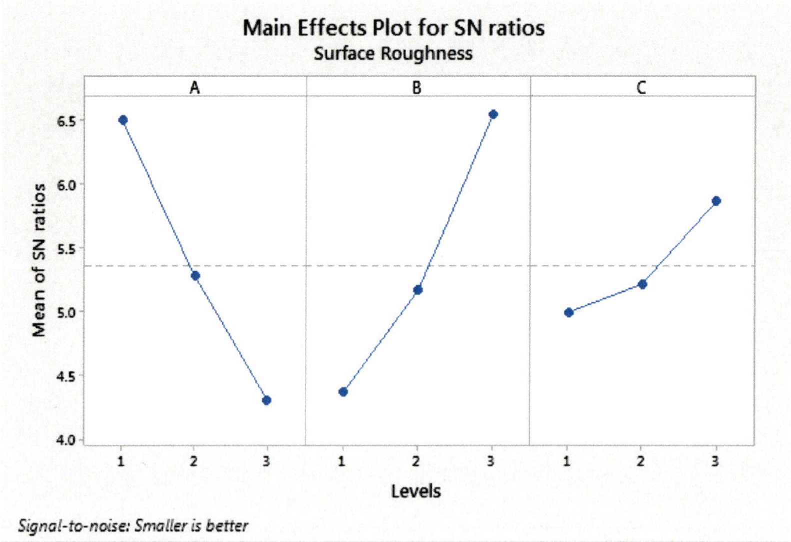

Fig. 4 (b). Main effect plot for SN ratios of SR.

Regression Analysis

The quantitative relationship between the input processing variables of the μ-ECM process as well as the required output parameters is focused mostly on results calculated from the experiments. Evidential connections that model the casting process are established through multiple linear regression. The influence of MRR and SR will be predicted using the below-mentioned equations. The correlation coefficients for the formulation of linear regression are 84.85% and 98.94% for MRR and SR, respectively.

$$MRR = -0.0133 + (0.00935 \times voltage) + (0.01827 \times electrolyte\ concentration) + (0.01827 \times current) \quad \textbf{(6)}$$

$$SR = -0.0900 + (0.06612 \times voltage) - (0.06533 \times electrolyte\ concentration) - (0.02292 \times current) \quad \textbf{(7)}$$

Multi-objective Optimization

The grey relation grades are calculated for each run and presented in Table **7**. The maximum value of grey relation grade is noted as rank 1. The optimum value found in run number 3 at voltage is 8 volts, electrolyte concentration is 0.40 mole/lit, and current is 1 amp.

Table 7. Grey relation grade and the experimental runs rank.

Run	Voltage (volt)	Electrolyte concentration (mole/lit)	Current (amps)	Grey relation grade	Rank
1	8	0.22	0.6	0.3812	8
2	8	0.31	0.8	0.4449	6
3	8	0.40	1	0.9656	1
4	12	0.22	0.8	0.3556	9
5	12	0.31	1	0.5788	4
6	12	0.40	0.6	0.6215	3
7	16	0.22	1	0.4693	5
8	16	0.31	0.6	0.3869	7
9	16	0.40	0.8	0.7245	2

Confirmation Test

The confirmation test is carried out for the optimum level using μ-ECM technique. Table **8** gives the values of the confirmation test, the predicted values of optimum run using regression models and indicates the percentage of deviation between the predicted and confirmation test results. From the study, the percent error is too low to imply that the Taguchi process is acceptable for measuring MRR and SR.

Table 8. Confirmation test results for an optimum level.

Condition	Run	Voltage (volt)	Electrolyte concentration (mole/lit)	Current (amps)	MRR (mm^3/min)	SR (μm)
Experimental	3	8	0.40	1	0.0887	0.3859
Prediction					0.0871	0.3899
% of error					1.84%	1.02%

Application

According to the results, the μ-ECM method is utilized for die sinking, profiles and contours, boring, polishing, and microscopic milling of the extremely hard material of Inconel 718 alloy. It is being used for machine steam turbine rotors under strict tolerances.

CONCLUSION

The present work is focussed on multi-objective optimization for Inconel 718 alloys in μ-ECM technique employing Grey Functional approach based on Taguchi. The results are obtained from the diverse discussions:

- The larger MRR is achieved at run number 9 (16 volts, 0.40 mole/lit, and 0.8 amp) of 0.0911 mm^3/min, and the better (lower) SR is obtained at run number 3 (8 volts, 0.40 mole/lit, and 1 amp) of 0.3854 μm.
- The voltage has been the most influencing variable for SR, and electrolyte concentration has been the most influencing input variable for MRR performance.
- From the main effect plots, it is identified that the MRR increases when voltage and electrolyte concentration increases. Moreover, there is a fluctuation in the response with respect to input process parameter current. The SR value increases when the input variables electrolyte concentration and current increases, yet the SR decreases when voltage is going down from 16 to 8 volts.
- The correlation coefficients for the developed regression equation for MRR and

SR are 84.85% and 98.94%, respectively. These regression models will be used for future predictions.

- From the grey relation technique, the optimum value was found in run number 3 and the voltage is 8 volts, electrolyte concentration is 0.40 mole/lit, and current is 1 amp.

This study aims to provide substantial support for manufacturing companies to increase efficiency and customer satisfaction in manufacturing Inconel 718 alloys using the μ-ECM technique.

CONSENT FOR PUBLICATION

Not applicable.

CONFLICT OF INTEREST

The author declares no conflict of interest, financial or otherwise.

ACKNOWLEDGEMENTS

Declared none.

REFERENCES

[1] C. Wang, X.P. Tan, S.B. Tor, and C.S. Lim, "Machine learning in additive manufacturing: State-o--the-art and perspectives", *Addit. Manuf.*, vol. 36, p. 101538, 2020. [http://dx.doi.org/10.1016/j.addma.2020.101538]

[2] F. Mehrabi, M. Farahnakian, S. Elhami, and M.R. Razfar, "Application of electrolyte injection to the electro-chemical discharge machining (ECDM) on the optical glass", *J. Mater. Process. Technol.*, vol. 255, pp. 665-672, 2018. [http://dx.doi.org/10.1016/j.jmatprotec.2018.01.016]

[3] X. Chen, Z. Wang, Y. Wang, and G. Chi, "Micro sinking electro-discharge machining of complex 3D micro-cavity using *in-situ* milled graphite microform electrode"., *J. Mater. Process. Technol.*, vol. 270, pp. 206-215, 2019. [http://dx.doi.org/10.1016/j.jmatprotec.2019.02.028]

[4] D.K. Rajak, D.D. Pagar, R. Kumar, and C.I. Pruncu, "Recent progress of reinforcement materials: a comprehensive overview of composite materials", *J. Mater. Res. Technol.*, vol. 8, no. 6, pp. 6354-6374, 2019. [http://dx.doi.org/10.1016/j.jmrt.2019.09.068]

[5] S. Rajesh, R. Suresh Kumar, S. Madhankumar, M. Sheshan, M. Vignesh, and R. Sanjay Kumar, "Study of the mechanical properties of Al7075 alloy, silicon carbide and fly ash composites manufactured by stir casting technique", *Mater. Today Proc.*, vol. 45, pp. 6438-6443, 2020. [http://dx.doi.org/10.1016/j.matpr.2020.11.278]

[6] S. Madhankumar, S. Rajesh, R. Balamurugan, N. Tharun Sri Ram, S. Venuprasath, and S. Tazmeel Ahamed, "Optimization of Material Removal Rate and Surface Roughness for Micro ECM of Inconel 718 alloy utilizing Grey Relational Technique", *IOP Conf. Series Mater. Sci. Eng.*, vol. 1059, no. 1, 2021. [http://dx.doi.org/10.1088/1757-899X/1059/1/012008]

[7] B. Bhattacharyya, B. Doloi, and P.S. Sridhar, "Electro-chemical micro-machining: new possibilities for micro-manufacturing", *J. Mater. Process. Technol.,* vol. 113, no. 1, pp. 301-305, 2001. [http://dx.doi.org/10.1016/S0924-0136(01)00629-X]

[8] S. Madhankumar, "Assessments and comparison of Inconel 625 and Inconel 718 alloys from overcut in micro ECM", In: *Materials Today.* vol. 45. Proceedings, 2020.

[9] D. Zhu, Y.B. Zeng, Z.Y. Xu, and X.Y. Zhang, "Precision machining of small holes by the hybrid process of electro-chemical removal and grinding", *CIRP Ann.,* vol. 60, no. 1, pp. 247-250, 2011. [http://dx.doi.org/10.1016/j.cirp.2011.03.130]

[10] P. Domanowski, and J. Kozak, "Direct and inverse problems of shaping by electro-chemical generating machining", *J. Mater. Process. Technol.,* vol. 107, no. 1, pp. 300-306, 2000. [http://dx.doi.org/10.1016/S0924-0136(00)00671-3]

[11] J. Song, W. Xu, Y. Lu, and X. Fan, "Rapid fabrication of superhydrophobic surfaces on copper substrates by electro-chemical machining", *Appl. Surf. Sci.,* vol. 257, no. 24, pp. 10910-10916, 2011. [http://dx.doi.org/10.1016/j.apsusc.2011.07.140]

[12] X. Zhu, Y. Liu, J. Zhang, K. Wang, and H. Kong, "Ultrasonic-assisted electrochemical drill-grinding of small holes with high-quality", *J. Adv. Res.,* vol. 23, pp. 151-161, 2020. [http://dx.doi.org/10.1016/j.jare.2020.02.010] [PMID: 32123588]

[13] S. Madhankumar, K. Manonmani, S. Periyasamy, and R. Suresh, "Experimental Study of Effect of Parameters on Material Removal Rate for Electrochemical Machining of Aluminium Matrix Composite", *Int. J. Creat. Res. Thoughts,* vol. 5, no. October, pp. 1062-1065, 2017.

[14] S. Hinduja, and M. Kunieda, "Modelling of ECM and EDM processes", *CIRP Ann.,* vol. 62, no. 2, pp. 775-797, 2013. [http://dx.doi.org/10.1016/j.cirp.2013.05.011]

[15] W. Han, and M. Kunieda, "Research on servo feed control of electrostatic induction feeding micro-ECM", *Precis. Eng.,* vol. 45, pp. 195-202, 2016. [http://dx.doi.org/10.1016/j.precisioneng.2016.02.012]

[16] M. Nakamura, and M. Kunieda, "Study of precision ECM in stationary electrolyte using stamp flushing method", *CIRP Ann.,* vol. 68, no. 1, pp. 173-176, 2019. [http://dx.doi.org/10.1016/j.cirp.2019.04.103]

[17] H. Luo, D. Mi, and W. Natsu, "Characteristics of ECM polishing influenced by work-piece corner feature and electrolyte flow", *Precis. Eng.,* vol. 56, pp. 330-342, 2019. [http://dx.doi.org/10.1016/j.precisioneng.2019.01.003]

[18] L.M. Maiyar, R. Ramanujam, K. Venkatesan, and J. Jerald, "Optimization of machining parameters for end milling of Inconel 718 super alloy using Taguchi based grey relational analysis", *Procedia Eng.,* vol. 64, pp. 1276-1282, 2013. [http://dx.doi.org/10.1016/j.proeng.2013.09.208]

[19] S. Madhankumar, R. Balamurugan, and S. Rajesh, "Investigations on austenitic nickel-chromium based super alloys - Inconel 625 and Inconel 718 from material removal rate in micro electro-chemical machining", *AIP Conf. Proc.,* vol. 2128, p. 040009, 2019. [http://dx.doi.org/10.1063/1.5117971]

[20] S. Madhankumar, "Experimental investigation on ultimate tensile strength of laser butt welded inconel 718 alloy and 2205 duplex stainless steel", In: *Materials Today.* vol. 45. Proceedings, 2020.

[21] A. Selvan, "T, S. Madhankumar, and K. Manonmani, "Optimization of Material Removal Rate and overcut in Micro Electrochemical Optimization of Material Removal Rate and overcut in Micro Electrochemical Machining of Inconel 718 using Taguchi Analysis", *Int. J. Pure Appl. Math.,* vol. 119, no. 12, pp. 869-877, 2018.

[22] S. Madhankumar, S. Rajesh, R. Balamurugan, T. Vignesh, and S. Jha, "Optimization of Process

Parameters on Surface Roughness for Micro Electrochemical Machining of Aluminium Matrix Composite", *Int. J. Sci. Res. Eng. Technol.,* vol. 7, no. 2, pp. 25-28, 2018.

[23] F. Klocke, M. Zeis, A. Klink, and D. Veselovac, "Experimental Research on the Electrochemical Machining of Modern Titanium- and Nickel-based Alloys for Aero Engine Components", *Procedia CIRP,* vol. 6, pp. 368-372, 2013. [http://dx.doi.org/10.1016/j.procir.2013.03.040]

[24] A.K.M. De Silva, H.S.J. Altena, and J.A. McGeough, "Influence of Electrolyte Concentration on Copying Accuracy of Precision-ECM", *CIRP Ann.,* vol. 52, no. 1, pp. 165-168, 2003. [http://dx.doi.org/10.1016/S0007-8506(07)60556-3]

[25] M. Burger, L. Koll, E.A. Werner, and A. Platz, "Electro-chemical machining characteristics and resulting surface quality of the nickel-base single-crystalline material LEK94", *J. Manuf. Process.,* vol. 14, no. 1, pp. 62-70, 2012. [http://dx.doi.org/10.1016/j.jmapro.2011.08.001]

[26] J. C. da S. Neto, E. M. da Silva, and M. B. da Silva, "Intervening variables in electro-chemical machining", *J. Mater. Process. Technol.,* vol. 179, no. 1, pp. 92-96, 2006.

[27] R.V. Rao, P.J. Pawar, and R. Shankar, "Multi-objective optimization of electro-chemical machining process parameters using a particle swarm optimization algorithm", *Proc. Inst. Mech. Eng., B J. Eng. Manuf.,* vol. 222, no. 8, pp. 949-958, 2008. [http://dx.doi.org/10.1243/09544054JEM1158]

[28] N. Manikandan, S. Kumanan, and C. Sathiyanarayanan, "Multi response optimization of electro-chemical drilling of titanium Ti6Al4V alloy using Taguchi based grey relational analysis", *Indian J. Eng. Mater. Sci.,* vol. 22, no. 2, pp. 153-160, 2015.

[29] S. Dhanabalan, K. Sivakumar, and C.S. Narayanan, "Analysis of form tolerances in electrical discharge machining process for inconel 718 and 625", *Mater. Manuf. Process.,* vol. 29, no. 3, pp. 253-259, 2014. [http://dx.doi.org/10.1080/10426914.2013.852213]

[30] S. Tripathy, and D.K. Tripathy, "Multi-attribute optimization of machining process parameters in powder mixed electro-discharge machining using TOPSIS and grey relational analysis", In: *Eng. Sci. Technol. an Int. J* vol. 19., 2016, pp. 62-70. [http://dx.doi.org/10.1016/j.jestch.2015.07.010]

[31] S. Madhankumar, R. Balamurugan, S. Rajesh, and K.M. Senthilkumar, "Fabrication of Al6063 alloy, silicon carbide and boron glass powder metal matrix composites in stir casting process and analysis the impact of process variables on mechanical properties", In: *Materials Today.* vol. 42. Proceedings, 2020.

[32] R.S. Veerapandian, S. Madhankumar, and S.H. Krishnan, "Experinmental Determination and Optimization of Hard Turned Austempered Ductile Iron (ADI) With AL-TI-N Coated Carbide Insert", *Int. J. Appl. Eng. Res.,* vol. 13, no. 11, pp. 9596-9600, 2018.

[33] S. Madhankumar, K. Manonmani, V. Karthickeyan, and N. Balaji, "Optimization of ultimate tensile strength of welded Inconel 625 and duplex 2205", *J. Mech. Eng. Sci.,* vol. 15, no. 1, pp. 7715-7728, 2021. [http://dx.doi.org/10.15282/jmes.15.1.2021.08.0608]

[34] S. Madhankumar, K. Manonmani, S. Rajesh, R. Balamurugan, and M. Harikrishnan, "Optimization of Micro Electrochemical Machining of Inconel 625 using Taguchi based Grey Relational Analysis", *Int. J. Appl. Eng. Res.,* vol. 13, no. 9, pp. 6771-6779, 2018.

[35] T.A. Selvan, S. Madhankumar, A. Viswanathan, and K.M. Senthilkumar, "Multi-objective optimization on output responses for micro ECM of Al6063 alloy, silicon carbide and boron glass powder metal matrix composites", In: *Materials Today.* vol. 45. Proceedings, 2020.

SUBJECT INDEX

A

B

C

C. Samson Jerold Samuel, M. Suresh & S. Gnanasekaran (Eds.)

D

E

N

O

P

R

S

Made in the USA
Las Vegas, NV
15 November 2026

08bfd25c-02cf-4970-9906-ae5948015bb4R04